ENERGY SCIENCE, ENGINEERING AND TECHNOLOGY

SOLAR PHOTOVOLTAICS

U.S. MANUFACTURING TRENDS AND TRADE

ENERGY SCIENCE, ENGINEERING AND TECHNOLOGY

Additional books in this series can be found on Nova's website under the Series tab.

Additional E-books in this series can be found on Nova's website under the E-book tab.

RENEWABLE ENERGY: RESEARCH, DEVELOPMENT AND POLICIES

Additional books in this series can be found on Nova's website under the Series tab.

Additional E-books in this series can be found on Nova's website under the E-book tab.

ENERGY SCIENCE, ENGINEERING AND TECHNOLOGY

SOLAR PHOTOVOLTAICS

U.S. MANUFACTURING TRENDS AND TRADE

JASON E. FOSTER
AND
ARTHUR M. POWELL
EDITORS

New York

For permission to use material from this book please contact us:
Telephone 631-231-7269; Fax 631-231-8175
Web Site: http://www.novapublishers.com

Library of Congress Cataloging-in-Publication Data

ISBN: 978-1-62257-234-2

Published by Nova Science Publishers, Inc. † New York

CONTENTS

PREFACE

In recent years, job creation and the development of a domestic renewable energy manufacturing base have joined national security and environmental concerns as rationales for promoting the manufacturing of solar power equipment in the United States. The most widely used solar technology involves photovoltaic (PV) solar modules, which draw on semiconducting materials to convert sunlight into electricity. By year-end 2011, the total number of grid-connected PV systems nationwide reached almost 215,000. Domestic demand is met both by imports and by about 100 U.S. manufacturing facilities employing an estimated 25,000 U.S. workers in 2011. Production is clustered in a few states, including, California, Oregon, Texas, and Ohio. This book discusses the solar photovoltaic industry and its supply chain; employment trends; international trade flows; and federal policy efforts aimed at supporting the industry.

Chapter 1 - Every president since Richard Nixon has sought to increase U.S. energy supply diversity. In recent years, job creation and the development of a domestic renewable energy manufacturing base have joined national security and environmental concerns as rationales for promoting the manufacturing of solar power equipment in the United States. The federal government maintains a variety of tax credits, loan guarantees, and targeted research and development programs to encourage the solar manufacturing sector, and state-level mandates that utilities obtain specified percentages of their electricity from renewable sources have bolstered demand for large solar projects.

Chapter 2 - Despite a sluggish U.S. economy in 2010, the U.S. photovoltaic (PV) industry saw strong growth. Total shipments of PV modules in 2010 more than doubled compared to total module shipments in 2009, rising from nearly 1.2 peak gigawatts (GWdc peak) to more than 2.6 peak gigawatts. This surge in growth was supported in part by rapid declines in the price of PV cells and modules, and by government incentives and policies at the Federal, state and local levels.

As a consequence of this strong growth, favorable business opportunities emerged for companies to expand or enter into the PV market. In 2010, there were 112 active PV manufacturers, importers and exporters that shipped PV cells and modules. This marks an increase of nearly 11 percent from the 101 companies doing business in 2009.

In: Solar Photovoltaics
Editors: J. E. Foster and A. M. Powell
ISBN: 978-1-62257-234-2

Chapter 1

U.S. SOLAR PHOTOVOLTAIC MANUFACTURING: INDUSTRY TRENDS, GLOBAL COMPETITION, FEDERAL SUPPORT*

Michaela D. Platzer

SUMMARY

Every president since Richard Nixon has sought to increase U.S. energy supply diversity. In recent years, job creation and the development of a domestic renewable energy manufacturing base have joined national security and environmental concerns as rationales for promoting the manufacturing of solar power equipment in the United States. The federal government maintains a variety of tax credits, loan guarantees, and targeted research and development programs to encourage the solar manufacturing sector, and state-level mandates that utilities obtain specified percentages of their electricity from renewable sources have bolstered demand for large solar projects.

---

* This is an edited, reformatted and augmented version of the Congressional Research Service Publication, CRS Report for Congress R42509, dated April 27, 2012.

The most widely used solar technology involves photovoltaic (PV) solar modules, which draw on semiconducting materials to convert sunlight into electricity. By year-end 2011, the total number of grid-connected PV systems nationwide reached almost 215,000. Domestic demand is met both by imports and by about 100 U.S. manufacturing facilities employing an estimated 25,000 U.S. workers in 2011. Production is clustered in a few states, including California, Oregon, Texas, and Ohio.

Domestic PV manufacturers operate in a dynamic and highly competitive global market now dominated by Chinese and Taiwanese companies. All major PV solar manufacturers maintain global sourcing strategies; the only U.S.-based manufacturer ranked among the top ten global cell producers in 2010 sourced the majority of its panels from its factory in Malaysia. Some PV manufacturers have expanded their operations beyond China to places like the Philippines and Mexico. Overcapacity has led to a significant drop in module prices, with solar panel prices falling more than 50% over the course of 2011. Several PV manufacturers have entered bankruptcy and others are reassessing their business models. Although hundreds of small companies are engaged in PV manufacturing around the world, profitability concerns appear to be driving consolidation, with ten firms now controlling half of global cell and module production.

The Department of Commerce and the U.S. International Trade Commission are investigating allegations that U.S. producers have been injured by dumped and subsidized imports from China. If significant duties are ultimately imposed, U.S. production could become more competitive with imports, but the cost of installing solar systems might rise. On the other hand, a number of federal policies that have helped to spur domestic demand for solar PV products have expired or reached their funding limits. These include the 1603 cash grant program and the advanced energy manufacturing tax credit; S. 591, which would extend the credit, has been introduced in the 112th Congress. Under current law, the Investment Tax Credit for PV systems will sunset at the end of 2016.

The competitiveness of solar PV as a source of electric generation in the United States will likely be adversely affected both by the expiration of these tax provisions and by the rapid development of shale gas, which has the potential to lower the cost of gas-fired power generation and reduce the cost-competitiveness of solar power, particularly as an energy source for utilities. In light of these developments, the ability to build a significant U.S. production base for PV equipment is in question.

INTRODUCTION

Major trends shaping the domestic photovoltaic (PV) manufacturing sector include technological advances, improved production methods, and a global surplus of manufacturing capacity,[1] especially from China. At the same time, PV manufacturers are grappling with falling module prices, which have adversely affected the operations of many solar companies, forcing some to reassess their business models and others to close factories or declare bankruptcy. Lower prices may be good for PV consumers, but they are squeezing manufacturers, especially in the United States and Europe. In addition, the rapid development of shale gas has the potential to lower the cost of gas-fired power generation in the United States, potentially affecting the competitiveness of solar power. In light of these trends, the ability to build a sustained U.S. production base for PV equipment is now in question.

U.S. solar manufacturing comprises a small part of the U.S. manufacturing base. In 2011, it directly employed about 25,000 workers, according to the Solar Energy Industries Association (SEIA), a trade group. [2] The U.S. cell and module market, measured by domestic shipment revenues, has grown in size from $1.2 billion in 2006 to $6.4 billion in 2010, reports the U.S. Energy Information Administration.[3] Following an unprecedented period of growth, the number of installed PV systems in the United States reached 214,157 by the end of 2011, more than twice the total at the end of 2009.[4]

Government support has been instrumental in sustaining the solar industry worldwide. In the United States, tax incentives and stimulus funding spurred recent double-digit growth rates in new PV installations.[5] Nevertheless, even with direct government involvement, solar energy still accounts for less than 0.1% of overall U.S. electricity generation.[6] The Obama Administration actively supports greater deployment of solar energy and sees it as one way to encourage advanced manufacturing in the United States, create skilled manufacturing jobs, and increase the role of renewable energy technology in energy production, among other objectives. In its *Blueprint for a Secure Energy Future*, the Obama Administration argues:

> We invented the photovoltaic solar panel, built the first megawatt solar power station, and installed the first megawatt-sized wind turbine. Yet today, China has moved passed us in wind capacity, while Germany leads the world in solar. To rise to this challenge, we need to tap into the greatest resource we have: American ingenuity.[7]

A PV Glossary

PV stands for photovoltaic derived from "photo" for light and "voltaic" for a volt, a unit of electrical force.

Solar photovoltaic, or solar PV for short, is a technology that uses the basic properties of semiconductor materials to transform solar energy into electrical power.

A *solar PV cell* is an electricity-producing device made of semiconducting materials. Cells come in many sizes and shapes. Materials used to make cells include monocrystalline silicon, polycrystalline silicon, amorphous silicon (a-Si), cadmium telluride (CdTe), copper indium gallium (dis)selendie (CIGS), and copper indium (di)selinide (CulnSe2 or CIS).

Panels, or *modules*, are comprised of a number of solar cells.

An array is the collective name for a number of *solar modules* connected together.

The anatomy of a solar cell, and how solar panels work, can be viewed at http://www.pbs.org/wgbh/nova/tech/how-solar-cell-works.html.

This report discusses the solar photovoltaic industry and its supply chain; employment trends; international trade flows; and federal policy efforts aimed at supporting the industry. It does not cover other methods of solar-power generation, such as concentrated solar power.[8] Concentrated solar technologies, largely dormant prior to 2006, are suitable mainly for utility-scale generation, whereas solar photovoltaics can be arranged in small-scale installations to produce power for individual buildings as well as in large installations to supply power to utilities.

One of the main federal policy tools to encourage solar generation is the federal solar investment tax credit (ITC)[9] for both residential and commercial solar installations, which is in effect until the end of 2016.[10] Stimulus funding in the American Recovery and Reinvestment Act of 2009 (ARRA)[11] included a U.S. Department of the Treasury grant in lieu of the ITC, the 1603 program, under which applicants through the end of 2011 received a 30% cash grant for eligible installed PV costs.[12] Other policy drivers include a federal loan guarantee program and the advanced manufacturing tax credit along with state renewable portfolio standards in more than half the states, mandating production of electricity from "clean" sources.[13] The SunShot initiative to advance domestic solar-based electricity generation includes various research and development (R&D) programs to strengthen PV manufacturing in the United States. No nationwide renewable electricity standard currently exists. However, the Obama Administration and some Members of Congress have endorsed the concept of a Clean Energy Standard, which would require utilities to purchase renewable energy.[14] While some of these policies do not directly address manufacturing, greater solar power adoption may support the development of a U.S. solar-energy manufacturing base.

Over the years, some European and Asian governments have enacted solar-promoting policies, including tax and electricity rate-payer subsidies, like feed-in tariffs (FITs), to spur their domestic markets.[15] Because of the recent economic crisis, European governments are beginning to eliminate, reduce, or change their incentive programs for solar power. The Japanese government has also sustained its domestic solar PV market by offering various inducements including a FIT, tax incentives, and direct grants for solar PV.[16] Elsewhere in Asia, countries such as China, Malaysia, and the Philippines provide various types of support to develop their domestic solar manufacturing sectors, which along with low labor costs, have made them hubs for solar PV production.

Even with decreasing PV prices, producing equipment that generates solar power at prices competitive with electricity generated from fossil fuels remains a challenge for manufacturers. This is especially true for utility-scale installations, as wholesale purchasers of electricity will compare the cost per megawatt hour of solar power directly with the cost of power from other sources. The cost-competitiveness of solar power is better in the residential and business markets, as the relevant comparison is with the delivered cost of electricity rather than with the generating cost. But even if the popularity of solar systems grows, falling equipment prices are likely to further challenge the profitability of manufacturers and interfere with efforts to sustain a solar manufacturing base in the United States.

SOLAR PHOTOVOLTAIC (PV) MANUFACTURING

Solar PV manufacturing, previously undertaken by numerous small firms, is rapidly maturing into a global industry dominated by a far smaller number of producers. Cell manufacturers typically have proprietary designs that seek to convert sunlight into electricity at the lowest total cost per kilowatt hour.

Vertical integration is becoming more important among the world's largest solar cell and module manufacturers, but many still rely on extensive supply chains for components such as wafers, glass, wires, and racks. Worldwide, the market for solar PV (including modules, system components, and installations) expanded from $2.5 billion in 2000 to $71.2 billion in 2010, according to one estimate, with the United States accounting for roughly 7%, or just over $5 billion, in 2010.[17]

Historical Overview

Modern photovoltaic technology traces its roots back to 19th-century breakthroughs by scientists from Europe and the United States. In 1839, a French physicist, Alexandre Edmond Becquerel, discovered the photovoltaic effect,[18] and in 1883, an American inventor, Charles Fritts, made the first primitive solar cell.[19] Progress in modern solar cell manufacturing began in the 1940s and 1950s when Russell Ohl discovered that a rod of silicon with impurities created an electric voltage when illuminated and three scientists at Bell Laboratories in New Jersey (Daryl Chapin, Calvin Fuller, and Gerald Pearson) developed the first commercial photovoltaic cell.

Further advancing PV cell manufacturing was the space race of the 1960s, with the competition between the United States and the former Soviet Union driving demand for solar cells, which were, and still are, used to power some spacecraft and satellites.[20] The first generation of photovoltaic manufacturing firms included such names as Hoffman Electronics, Heliotek,[21] RCA, International Rectifier, and Texas Instruments. The technology, however, remained too expensive for other uses, and the market remained very small.[22] The Japanese manufacturer Sharp pioneered the use of photovoltaics on earth, using them to power hundreds of lighthouses along the Japanese coast, but it could not identify other applications for which photovoltaics were cost-competitive.

The oil crises of the 1970s hastened the development of modern solar panels by a second generation of PV firms, which focused on ground applications. Major oil and gas companies entered the field.[23] Exxon underwrote the Solar Power Corporation.[24] Atlantic Richfield Company (ARCO) purchased Solar Technology International and renamed it ARCO Solar in 1977; its corporate descendant is now part of Solarworld, presently the largest cell manufacturer in the United States. First Solar, one of the biggest manufacturers of PV thin-film cells, can trace its roots to Toledo, OH, where it was established in 1984 as Glasstech Solar.

The first direct federal support for solar manufacturing was during the Carter Administration. The Energy Tax Act (ETA) of 1978[25] provided tax credits for homeowners who invested in solar and certain other technologies. Additionally, the federal government through the Public Utility Regulatory Policies Act required utilities to purchase power produced by qualified renewable power facilities.[26]

Notwithstanding this support, production of solar PV power in the United States remained small. By the mid-1980s, domestic photovoltaic manufacturers were selling products at a loss and many were struggling. President Reagan's Tax Reform Act of 1986 reduced the Investment Tax Credit (ITC) to 10% in 1988, where it remained until 2005. Because of these policy changes, combined with the sustained drop in petroleum prices, solar manufacturing slumped until 2005, when President George W. Bush signed the Energy Policy Act (EPAct).[27] That law included a 30% ITC for property owners who installed commercial and residential solar energy systems.[28]

The Manufacturing Process

PV systems do not require complex machinery and thousands of parts. In fact, most PV systems have no moving parts at all. They also have long service lifetimes, typically ranging from 10 to 30 years, with some minor performance degradation over time. In addition, PV systems are modular; to build a system to generate large amounts of power, the manufacturer essentially joins together more components than required for a smaller system. These characteristics make PV manufacturing quite different from production of most other types of generating equipment. In particular, PV systems offer little opportunity for manufacturers to make customized, higher-value products to meet unique needs. Manufacturers offer competing technological approaches to turning sunlight into electricity, but many customers have no reason to care about the technology so long as the system generates the promised amount of electricity. Economies of scale are significant, as increasing output tends to lower a factory's unit costs.

A technology known as crystalline silicon PV accounts for roughly 80% to 85% of global PV production capacity.[29] Production of a crystalline silicon system involves several stages:

- **Polysilicon manufacturing.** Polysilicon, based on sand, is the material used to make the semiconductors that convert sunlight into electricity. Its production requires large processing plants, with the construction of a polysilicon plant taking about two years and costing between $500 million and $1 billion.[30] Polysilicon comprises about a quarter of the cost of a finished solar panel.[31] Historically, polysilicon prices have been volatile, because the construction of a new plant can add a large amount of supply to the market. High polysilicon prices can adversely affect the profitability of manufacturers further down the supply chain. A handful of manufacturers from the United States, Europe, and Japan currently dominate polysilicon production, with much of it now located in Europe and the United States,[32] but increasingly manufacturers like GLC Solar from China and OCI from South Korea have expanded their production levels.
- **Wafer manufacturing.** Using traditional semiconductor manufacturing equipment, wafer manufacturers, including companies such as Sumco, Siltronic, Nexolon, and MEMC, shape polysilicon into ingots and then slice the ingots into thin wafers. The wafers are then cut, cleaned, and coated according to the specifications of the system manufacturers.
- **Cell manufacturing.** Solar cells are the basic building blocks of a PV system. They are made by cutting wafers into desired dimensions (typically 5 x 5 or 6 x 6inches) and shapes (round, square, or long and narrow). The manufacturer then attaches copper leads so the cell can be linked to other cells. Minimizing the area covered by these leads is a key issue in cell design, as the lead blocks sunlight from reaching parts of the cell surface and thus reduces potential energy output. The U.S. Department of Energy estimates that a manufacturing plant to produce 120 MW of cells per year would require an investment of around $40 million.[33]
- **Module manufacturing.** Modules, which normally weigh 34 to 62 pounds, are created by mounting 60 to 72 cells on a plastic backing within a frame, usually made of aluminum.[34] The module is covered by solar glass to protect against the elements and to maximize the efficiency with which the unit coverts sunlight into power. Production of solar glass is highly capital intensive, and approximately60% of the global market is controlled by four global manufacturers: Ashai, NSG Group (Pilkington), Saint Gobain, and Guardian.[35] The glass is expensive to ship, so glass producers tend to locate near module

manufacturers.[36] In some countries, module manufacturing is highly automated; in others, more labor-intensive processes are used.

A newer technology, thin-film PV, accounts for 10-15% of global installed PV capacity.[37] Rather than using polysilicon, these cells use thin layers of semiconductor materials like amorphous silicon (a-Si), copper indium diselenide (CIS), copper indium gallium diselenide (CIGS), or cadmium telluride (CdTe). The manufacturing methods are similar to those used in producing flat panel displays for computer monitors, mobile phones, and televisions: a thin photoactive film is deposited on a substrate, which can be either glass or a transparent film. Afterwards, the film is structured into cells. Unlike crystalline modules, thin-film modules are manufactured in a single step. Thin-film systems are usually less costly to produce than crystalline silicon systems, but have substantially lower efficiency rates.[38] On average, thin-film cells convert 5%-13% of incoming sunlight into electricity, compared to 11%-20% for crystalline silicon cells. However, as thin film is relatively new, it may offer greater opportunities for technological improvement.[39]

Crystalline silicon systems and thin-film systems all make use of a variety of other components, known as "balance of system" equipment. These include batteries (used to store solar energy for use when the sun is not shining), charge controllers, circuit breakers, meters, switch gear, mounting hardware, power-conditioning equipment, and wiring. In the United States, inverters are also needed to convert the electricity generated from direct current (DC) to alternating current (AC). Typically, balance of system components are not made by the system manufacturers, but are sourced from external suppliers.

Similar to many other advanced manufacturing industries, solar panel manufacturing depends on a global supply chain (see **Figure 1** for an overview), with PV manufacturers sourcing products at each stage of the value chain from suppliers located anywhere in the world. For instance, PV manufacturers purchase the majority of their solar factory equipment for wafer, cell, and module production from European and U.S. firms such as Roth & Rau (Germany), Applied Materials (United States), GT Solar (United States), and Oerlikon Solar (Switzerland). According to an analysis by Bloomberg New Energy Finance, a system produced by the U.S.-based firm SunPower may use polysilicon from a Korean supplier, DC Chemical; wafers from a First Philec-SunPower joint venture in the Philippines; cells manufactured at a SunPower factory in the Philippines; and modules assembled in Mexico or Poland.[40]

Each solar panel assembler uses different sourcing strategies, and the levels of vertical integration vary across the industry. At one extreme, SolarWorld, based in Germany, is highly integrated, controlling every stage from the raw material silicon to delivery of a utility-scale solar power plant. At the other extreme, some large manufacturers are pure-play cell companies, purchasing polysilicon wafers from outside vendors and selling most or all of their production to module assemblers. A number of solar manufacturers seem to be moving toward greater vertical integration for better control of the entire manufacturing process. Vertical integration also reduces the risk of bottlenecks holding up delivery of the final product.

Overall, labor accounts for about 10% of production costs in the industry, with module assembly accounting for a majority of labor costs in the production process.[41] Most stages of production are highly automated. A recent study by the U.S. International Trade Commission (ITC) reported that even the more labor-intensive module assembly process is being automated, and that module assembly in China and the United States uses similar levels of automation.[42] International transport costs for finished modules are also small, in the range of 1%-3% of value, producers told the ITC.[43]

Production and transportation costs, therefore, do not appear to be the major considerations determining where manufacturing facilities are located. For example, according to a National Renewable Energy Laboratory presentation, Chinese producers have an inherent cost advantage of no greater than 1% compared with U.S. producers; in the U.S. market, China suffers a 5% cost disadvantage when shipping costs are included.[44]

Production Locations

With neither labor costs nor transportation costs being decisive, many manufacturers that opened new facilities over the past decade chose to locate them in countries with strong demand—which is to say, in countries with attractive incentives for PV installations. Worldwide, the biggest markets have been Europe (principally Germany, Italy, and Spain) and Japan. Together, they comprised about two-thirds of the world's cumulative PV installed capacity of nearly 70 GW in 2011.[45] In Europe, until recently, government policies have fueled demand through such policy mechanisms as feed-in tariffs, which require utilities to purchase renewable power at generous rates, effectively forcing consumers to subsidize solar power through their electric bills.

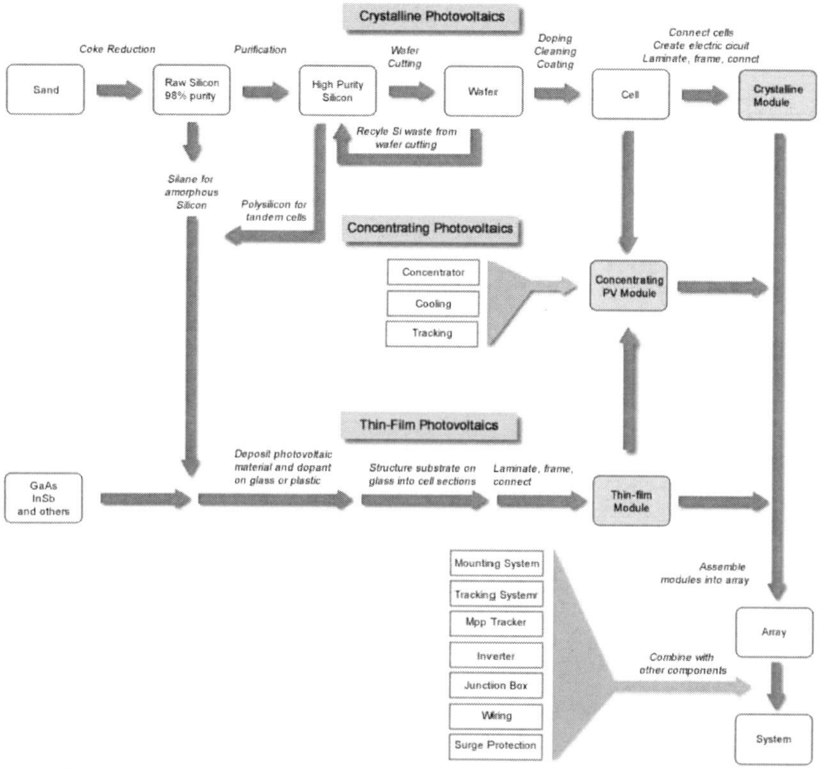

Source: Green Rhino Energy, http://www.greenrhinoenergy.com/solar/technologies/pv_valuechain.php. Reproduced with permission from Green Rhino Energy.

Figure 1. PV Value Chain.

The U.S. market for PV products is relatively small, accounting for about 7% of global PV installations in 2011, but has been growing at a rapid rate (see **Figure 2**).[46] The amount of solar capacity installed during 2011 was more than twice the 2010 amount.[47] The Solar Energy Industries Association reports that at year-end 2011, cumulative PV capacity in the United States reached almost 4 GW. Of new installations linked to the electric grid during 2011,

- 43% were for commercial or other non-residential customers, excluding utilities;
- 41% consisted of utility-scale installations, which generally use the largest panels and provide electricity directly to the electric grid; and
- 16%, the smallest share, were for residential buildings.[48]

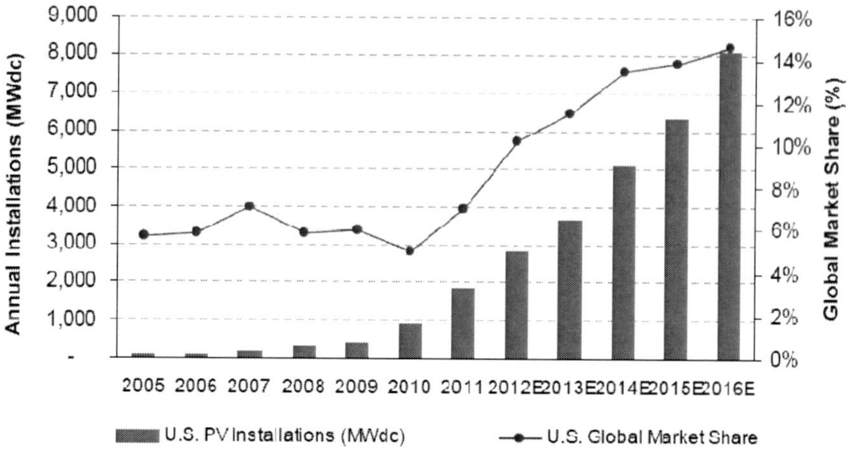

Source: SEIA/GTM Research "U.S. Solar Market Insight: 2011 Year in Review."

Notes: The annual installed figures cover only grid-connected capacity. DC stands for direct current, the type of power output by photovoltaic cells and modules.

Figure 2. U.S. PV Installations and Global Market Share.

Domestic Production

In the United States, manufacturers produced PV modules with a capacity of 1.1 peak gigawatts[49] (GW) in 2010, according to the Energy Information Administration.[50] By value, combined U.S. PV cell and module shipments totaled about $6.4 billion in 2010.[51] As shown in Table 1, three firms, SolarWorld, First Solar, and Suniva, accounted for nearly 60% of total domestic cell production.

The domestic solar manufacturing sector comprises about 100 production facilities making primary PV components (polysilicon, wafers, cells, modules, and inverters) as reported by SEIA.[52] SolarWorld's Oregon facility is the largest solar cell and module plant in the United States, with the capacity to produce 500 megawatts (MW) of solar cells per year at full production.[53] A number of other foreign-based firms, such as Schott Solar, Sanyo, Kyocera, and Siemens, operate domestic PV primary component plants, and China-based Suntech, the world's largest cell and module manufacturer, has a small plant in Arizona.[54]

As shown in **Figure 3**, manufacturing facilities for primary solar PV equipment and components are located throughout the United States, with concentrations of facilities in California, Oregon, Arizona, Ohio, Texas, and

Colorado. As noted above, due to the global supply chains prevalent in the PV industry, the amount of domestic content may vary considerably from one plant to another. The map does not include announced facilities that have yet to start operating.

A closer examination of SEIA's data shows that in 2011, eight U.S. facilities produced raw materials for the PV industry, including Hemlock, MEMC, and REC. Seven facilities are involved in wafer/ingot production. About another 50 facilities made cells or assembled modules, and 28 were involved in the production of solar inverters. SEIA's list does not include other parts of the PV supply chain, such as equipment for the PV industry or other balance of system components.

Table 1. Cell and Module Production in the United States in MW, 2010

Company	Location of Headquarters	Technology	Cells	Modules	% of U.S. Cell Production
SolarWorld	Germany	Mono/Multi c-Si	251	219	22.9%
First Solar	United States	CdTe	222	222	20.2%
Suniva	United States	Mono c-Si	170	15	15.5%
Evergreen Solar	United States	Mono/Multi c-Si	158	158	14.4%
United Solar	United States	a-Si	120	120	10.9%
Solyndra	United States	CIGS	67	67	6.1%
Solar Power Industries	United States	Mono/Multi c-Si	35	31	3.2%
Abound Solar	United States	CdTe	31	31	2.8%
Miasole	United States	CIGS	20	20	1.8%
Global Solar	United States	CIGS	17	0	1.5%
All Others			7	382	0.6%
Total			**1,098**	**1,265**	**100.0%**

Source: International Energy Agency, U.S. PV Applications National Survey Report, 2010, May 201, pp. 17-18.

Notes: C-Si stands for crystalline silicon. Monocrystalline PV cells are usually cut from a single grown silicon ingot, while multicrystalline PV cells are manufactured such that wafers are made from multiple crystals. Monocrystalline PV cells have an efficiency of 16% to almost 20%, while the cheaper to produce mutlicrystalline PV cells achieve an efficiency of 14% to 15%. Thin-film PV is based on other materials such as amorphous silicon (a-Si), cadmium telluride (cdTe), or copper iridium di-selenide (CIGS).

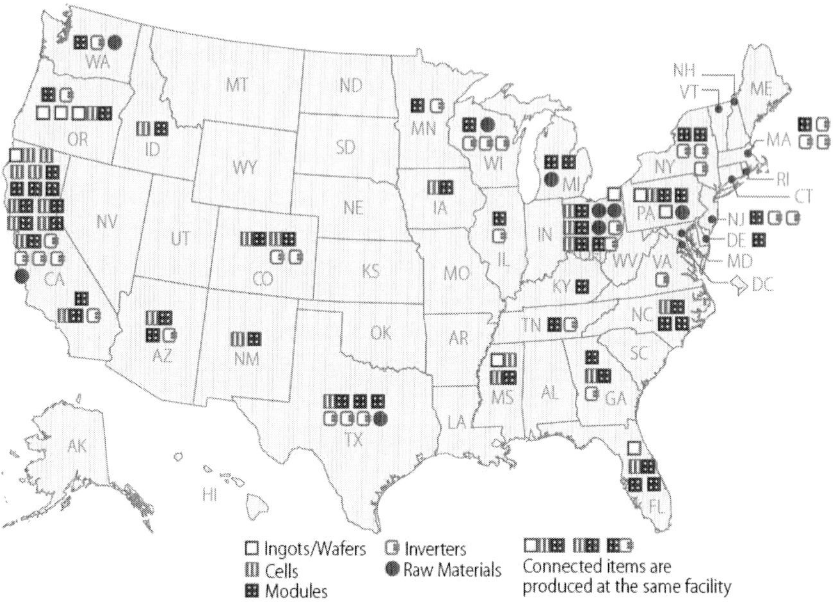

Source: Data provided to CRS by SEIA.

Notes: This map is not inclusive of all PV facilities in the United States.

Figure 3. U.S. Cell/Module and Polysilicon Production Facilities 2011.

PV production facilities appear to have relatively short life spans, at least in the United States. Industry data indicate that at least eight U.S. solar manufacturing facilities were closed in 2011. Of these, five had operated for less than five years. **Table 2** list some recent PV facility closures.

While some manufacturers have closed their U.S. facilities, others continue to open new U.S. manufacturing plants or expand existing ones.[55] SEIA's analysis of forthcoming PV manufacturing facilities notes, "there is a healthy spread across the value chain and technologies when it comes to new PV plants in the United States."[56] Future plants include a polysilicon facility (Calisolar) in Mississippi and a wafer manufacturing plant (1366 Technologies) in Massachusetts. GE Energy is building a $600 million 400 MW state-of-the-art thin-film CdTe manufacturing plant in Colorado.[57] Stion, a CIGS thin-film manufacturer, opened a new factory in Mississippi in 2011[58] and began commercial shipments in early 2012.[59] **Table 3** provides selected examples of U.S. PV manufacturing plants that could commence operations by 2014.

Table 2. Selected Recent PV Facility Closures

Company	Status	Year Online	Year Closed	State	Products
Evergreen Solar, Inc.	Closed	2008	2011	MA	Wafers
MEMC Southwest, Inc.	Closed	1995	2011	TX	Ingots
SolarWorld Americas[a]	Closed	2007	2011	CA	Modules
Solon America Corp.	Closed	2008	2011	AZ	Modules
Solar Power Industries	Closed	2003	2011	PA	Cells, modules
Solyndra, Inc	Closed	2010	2011	CA	Modules
SpectraWatt, Inc.[b]	Closed	2009	2011	NY	Cells
BP Solar[c]	Closed	1998	2012	MD	Cells, modules
Energy Conversion	Suspension of all	2003	2011	MI	Cells, modules
Devices	factories/sale pending				
Sanyo	Closed one factory	2003	2012	CA	Wafers

Source: SEIA. Annual Market Reports, 2010 and 2011.

a. SolarWorld purchased the California facility from Royal Dutch Shell in 2006 and expanded it with a $30 million investment. It remains open for sales and marketing activities, but production was moved to Oregon.

b. SpectraWatt was a 2008 spinoff from an internal research project by the Intel Corporation. The company began shipments from its New York facility in 2010.

c. Plant originally owned by Solarex, which opened it in 1981. In 1995, Amoco/Enron acquired Solarex and subsequently BP acquired it. In 2005, BP announced plans to double the plant's capacity.

U.S. Solar Manufacturing Employment

As shown in **Figure 4**, the solar manufacturing sector supported about 25,000 jobs nationwide in 2011, according to SEIA. This accounted for only about one-fourth of U.S. employment related to the solar energy sector.[60] The remaining 75% of the 100,000 full-time workers employed directly in the solar power industry as of August 2011 are involved in other segments of the industry, including installation, sales and distribution, project development, research and development, and finance.[61]

Table 3. Selected New or Planned PV Plants

Company	Status	Date Online	State	Product
1366 Technologies, Inc.	Planned	2013	MA	Wafers
Abound Solar[a]	Planned	2013/2014	IN	Module
Calisolar, Inc.	Planned	2013	MS	Raw Materials
First Solar, Inc.	Construction stopped	2012	AZ	Modules
Fronius USA, LLC	Planned	2012/2016	IN	PV - Inverters
GE Energy	Planned	2012	CO	Modules
Hemlock Semiconductor Corp.	Planned	2012	TN	Raw Materials
SoloPower	Planned	2012	OR	Module
Wacker Polysilicon	Under construction	2013	TN	Raw Materials

Source: SEIA. Annual Market Report, 2011

a. Abound Solar has announced "temporarily eliminating 180 full-time jobs" at its Colorado plant, and plans for its Tipton, IN plant now appear uncertain. See Abound Solar Production Plan FAQ at http://www.abound.com/feb28faq.

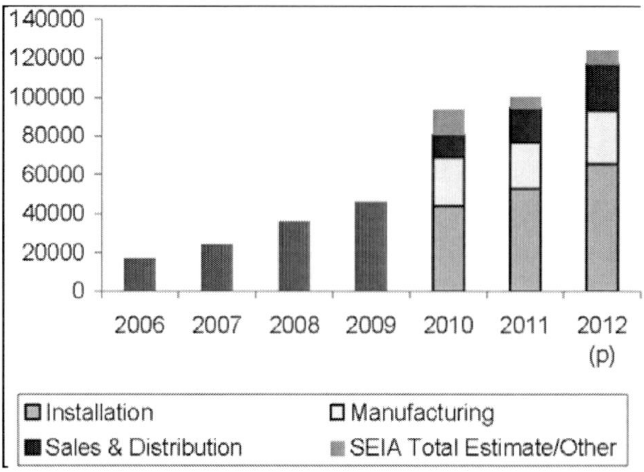

Source: SEIA, National Solar Job Census, 2011. 2012 data are preliminary

Notes: Other refers to project development, R&D, and finance. From 2006 to 2009, SEIA estimated the number of jobs and did not conduct a census for those years.

Figure 4. Domestic Solar Industry Employment Trends 2006-2012.

The number of solar manufacturing jobs has been relatively flat in recent years, even as total employment in the solar energy industry increased, according to figures from SEIA.[62] This is not surprising, as the majority of PV cells and modules are made overseas, including many that are manufactured by U.S. companies at offshore facilities. The near-term prospects for increased employment in solar manufacturing seem limited, as job creation from the opening of new plants may be outweighed by the jobs lost due to plant closures.

Solar manufacturing is responsible for a very small share of the 11.7 million domestic manufacturing jobs in 2011, well under 1%. Even given a substantial increase in U.S. solar manufacturing capacity, that solar PV manufacturing seems unlikely to become a major source of jobs. Employment growth is likely to depend not only upon future demand for solar energy, but also on corporate decisions about where to produce solar PV products, including components like inverters and other balance of system parts.

GLOBAL PRODUCTION SHIFTS

Recent policy actions by governments in a number of countries, including Germany, Italy, and the United States, indicate that energy consumers will have smaller incentives to install solar PV systems than in the recent past.[63] This may lessen the industry's eagerness to maintain production locations in many different countries. At the same time, due to technological developments and falling prices for polysilicon, the cost of solar cells and modules has been falling steeply (see **Figure 5**).[64] SolarBuzz, a market research firm, forecasts that over the next five years module prices will drop another 43-53% from 2011 levels.[65] Price pressures have driven a number of manufacturers, including the U.S. firms Evergreen Solar and Solyndra and the German companies Solon and Q-Cells, into bankruptcy, and have led others to lay off workers.

The creation of incentives for solar installations in several countries around 2004 led many companies to enter the PV industry. According to an estimate by Photon International, more than 1,000 PV module manufacturers worldwide supplied the market in 2011.[66] But with demand in some countries declining and prices weak, the industry appears to have entered a phase of rapid consolidation on a global basis. Meanwhile, some manufacturers in China and Taiwan continue to expand rapidly to obtain economies of scale and reduce unit costs (see **Figure 6**), potentially contributing to global overcapacity in PV production.

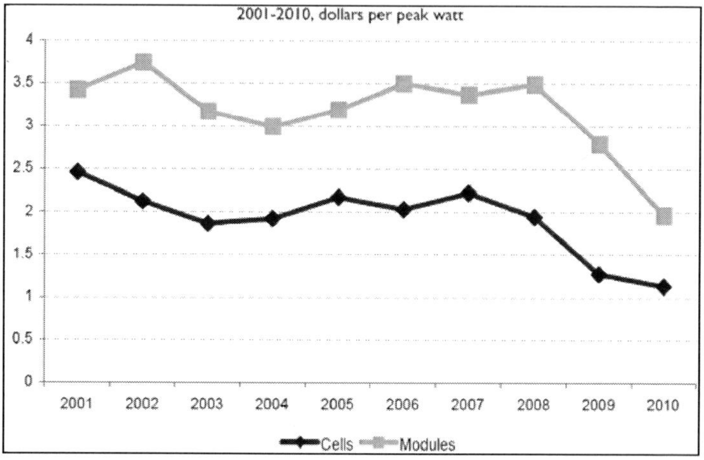

Source: U.S. Energy Information Administration, Form EIA-63B, "Annual Photovoltaic Cell/Module Shipments Report

Figure 5. Average Price of PV Cells and Modules.

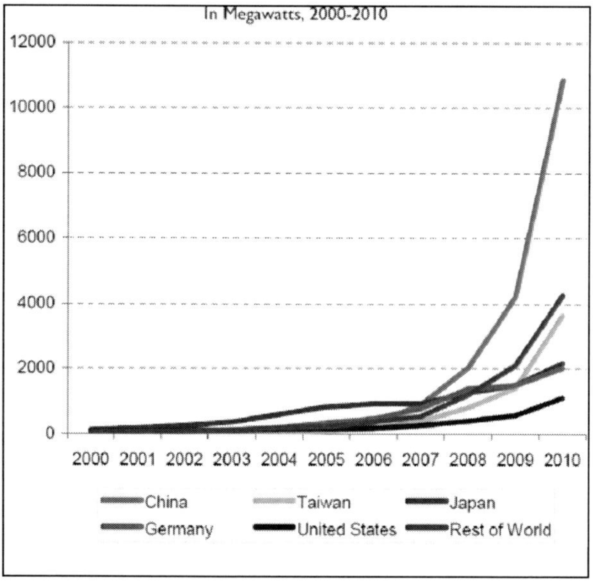

Source: Data compiled by the Earth Policy Institute from GTM Research, http://www.earth-policy.org/indicators/C47.

Figure 6. Annual Solar Cell Production by Country.

China currently exports about 95% of all the PV modules it produces.[67] Its domestic market for solar PV installations was small at less than 1 GW in total installed PV capacity in 2010. However, China has begun to implement policies to expand domestic solar PV demand, including direct grants for solar PV installations (close to $3 per watt for systems over 50 kW capacity).[68] More recently, it implemented a nationwide feed-in tariff.[69] Because of these policies, China's solar market may grow quickly, with SEIA forecasting that by 2016 it will be one of the world's leading markets by PV installations. By the end of 2011, cumulative installed and connected capacity in China had risen substantially to 2.9 GW.[70] The Indian market also may experience strong growth if the country aggressively implements its National Solar Mission, which aims to expand its domestic solar market to 20 GW of electricity by 2020.[71]

There is no dominant player in what is still a highly fragmented industry; more than 100 solar cell and more than 300 solar module companies are reported to exist in China alone.[72] But as some manufacturers have expanded and others have exited, ten firms now control about half of global production. Of these, four are based in China and two in Taiwan (see **Table 4**). All, however, are pursuing global business strategies.

U.S. TRADE IN SOLAR PRODUCTS

As part of their global business strategies, U.S. solar panel equipment manufacturers source a significant share of components outside the United States. Imports of solar cells and panels grew to nearly $5 billion by 2011 from just $227 million in 2005 (see **Table 5**).[73] PV imports have been rising for several reasons: (1) increasing crystalline silicon (c-Si) module production in places like China, Malaysia, and the Philippines; (2) an emergent U.S. market, responding to the falling price of solar energy; and (3) favorable state polices in key markets like California.[74] Solar cell imports are also rising because more European- and Asian-based firms have established crystalline module assembly plants in the United States. Some of the cells assembled at these U.S. assembly plants come from these companies' facilities overseas.

Two-thirds of solar cells and modules imported into the United States come from Asia. Topping the list is China, at $2.8 billion, accounting for 56% of all PV imports into the United States in 2011. China's lead is recent since most of its large PV manufacturers are young companies established over the last decade.[75] Malaysia is another large supplier of PV modules to the United

States, reflecting the greater production capacity of two U.S. companies, First Solar and AUO-SunPower, and the German producer, Q-Cells.

Table 4. Top PV Cell Manufacturers by Production 2010

Rank	Manufacturer	Location of Headquarters	% of Global Cell Production	Founded	Plant Locations (current and planned)
1	Suntech	China	6.6	2001	China, Japan, United States
2	JA Solar	China	6.1	2005	China
3	First Solar[a]	United States	5.9	1990	United States, Malaysia, Germany
4	Yingli Green Energy	China	4.7	1998	China
5	Trina Solar	China	4.7	1997	China
6	Q-Cells[b]	Germany	3.9	1999	Germany, Malaysia, Sweden
7	Gintech	Taiwan	3.3	2005	Taiwan
8	Sharp	Japan	3.1	1959	Japan, Italy, United States, UK, Thailand
9	Motech	Taiwan	3.0	1981	Taiwan and China
10	Kyocera	Japan	2.7	1996	Japan, Czech Republic, United States
11	Hanwha Solar	South Korea	2.2	2004	China, South Korea

Source: U.S. Department of Energy, 2010 Renewable Energy Databook. All other manufacturers accounted for 53.7% of global cell production in 2010.

a. In April 2012, First Solar announced it would close its manufacturing operations in Germany by the end of 2012, indefinitely idle some of its production lines in Malaysia, and ultimately reduce its global workforce by about 2,000 positions, or about 30% of the total. See First Solar April 17, 2012 press release, "First Solar Restructures Operations to Align with Sustainable Market Opportunities," for more information, http://investor.firstsolar.com/releasedetail.cfm?ReleaseID=664717.

b. In April 2012, Q-Cells announced that it would begin bankruptcy proceedings. For more information see,Q-Cells, "Q-Cells SE Filed for Insolvency Proceedings," April 3, 2012, http://www.q-cells.com/en/press/article//Q-Cells-SE-filed-for-insolvency-proceedings.html,

Table 5. U.S. Imports of Solar Cells and Modules, Select Countries in U.S. dollars, by selected years

Country	2005	2008	2010	2011	% Change, 2005-2011	% Change, 2010-2011
China	$22,185,547	$229,281,465	$1,192,336,468	$2,802,334,973	12,531%	135%
Malaysia	$177,539	$19,465	$139,098,366	$562,810,729	316,907%	305%
Mexico	$50,974,121	$213,202,533	$481,120,256	$514,335,119	909%	7%
Japan	$122,436,113	$250,938,688	$301,265,837	$392,681,769	221%	30%
Philippines	$645,673	$138,593,374	$27,891,274	$241,912,389	37,367%	767%
World	$227,143,964	$1,240,029,288	$2,644,989,618	$4,975,159,406	2,090%	88%

Source: Global Trade Atlas. These statistics only cover solar cells and panels (HS 8541406020 and HS 8541406030)

Notes: Imports are shown by domestic consumption.

Until 2008, Japan was the top exporter of solar panels and cells to the United States. By 2011, it dropped to the fourth-largest PV exporter, at $393 million. PV exports from the Philippines amounted to $242 million in 2011, largely due to SunPower's large production facility, where it does most of its manufacturing.[76] Because of investments by foreign PV manufacturers like Kyocera and Sanyo, which assemble PV modules in Mexico for export, U.S. imports of PV cells and modules from Mexico have grown, although they still remain small at just over $500 million in 2011.[77] U.S. imports of PV products from South Korea are small, but the country has a stated goal to capture 10% of the global PV market by 2020.[78]

Allegations of Dumped and Subsidized Solar PV Products from China

In October 2011, the Coalition for American Solar Manufacturing (CASM), led by the U.S. unit of SolarWorld, along with MX Solar US, Helios Solar Works, and four unnamed companies,[79] filed anti-dumping and countervailing duty petitions with the U.S. Department of Commerce (DOC) and the International Trade Commission (ITC), alleging that Chinese makers of crystalline silicon photovoltaic cells and modules have injured U.S. producers by selling their products in the United States at below-market prices.[80] The CASM petition asked the Commerce Department to levy tariffs of up to 250% on solar cells and modules imported from China. In a preliminary decision in March 2012, the department announced the imposition of modest tariffs of less than 5% on Chinese solar cells and modules.[81] A final determination is expected to come later in 2012.

The Coalition for Affordable Solar Energy (CASE)[82] opposes the CASM petition, claiming that higher tariffs on PV imports from China would curb domestic demand for solar products, could erode profit margins across the PV value chain, and might make it even harder for solar energy to compete with fossil fuels. Another claim by CASE is that a 100% tariff or above could cost the United States as many as 50,000 net jobs by 2014.[83] Chinese manufacturers have also called on their own Commerce Ministry to initiate an investigation into alleged U.S. subsidies and dumping of polysilicon exports to China, although such practices, if they are occurring, would lower the cost of producing finished cells and modules in China.

If the dumping and subsidy cases lead to significant duties against imports from China, Chinese solar cell and module manufacturers might attempt to shift production to other locations, such as South Korea, Taiwan, and the European Union, where the duties would not apply. Some Chinese producers may seek to avoid the duties by opening production in the United States.

Domestic Content

One estimate indicates that in 2010 U.S. content accounted for 20% of the value of U.S.-installed crystalline silicon modules and 71% of the value of U.S.-installed thin-film modules. These figures were slightly lower than the 2009 approximations on domestic content of U.S.-installed crystalline silicon modules and thin-film modules at 24% and 77%, respectively.[84] SEIA notes that there is "nothing intrinsically American about thin film manufacturing, intrinsically foreign about crystalline silicon production." It ascribes the higher U.S. value added in thin film to the fact that U.S. manufacturers like First Solar lead in thin-film production and that the sample size for thin-film manufacturers is small.

Estimates on the level of U.S.-sourced content for other segments of the PV industry include inverters, with domestic value increasing from 26% in 2009 to 45% in 2010; mounting structures up from 84% in 2009 to 94% in 2010; and combiner boxes and miscellaneous electrical equipment share of domestic value down from 61% in 2009 to 59% in 2010.[85] It is not possible to determine precisely the value of PV components created domestically and how much is imported because of the complex nature of the solar supply chain.

U.S. Exports

U.S. PV exports to the world remain relatively small at slightly more than $1 billion in 2011, but more than double the $442.7 million in 2006, according to data compiled from Global Trade Atlas. The ITC attributes U.S. export expansion to growing overseas markets, an expanding domestic industry, and a strategy of diversification.[86] In 2011, Canada and Germany were the two largest foreign markets for U.S. solar PV exports at $285 million and $207 million, respectively. The larger European Union market accounts for the majority of U.S. PV exports. There are essentially no PV module exports from the United States to China.

U.S. exporters of solar cells and panels generally do not face foreign tariffs because of the plurilateral Information Technology Agreement (ITA), whose signatories have agreed to eliminate duties on information technology products.[87] Tariffs in other parts of the PV value chain are also comparably low. For example, the applied tariff on silicon is between zero and 4% in the leading cell and module producing countries.[88] However, non-tariff barriers can be significant, including local content requirements at the national level or sub-national level in places like India and Canada and other policies that encourage the use of local content in countries like Italy. Besides these mandates, import charges and taxes, customs procedures, and divergent product standards can hinder trade in solar PV components.[89] Subsidies for domestic production in major overseas markets like China are another potential constraint on U.S. exports.[90]

Several U.S. government programs encourage the export of renewable energy products. Targeting large emerging markets like India, the Export-Import Bank provides direct loans to solar manufacturers through its Environmental Products Program, under which it allocates a certain portion of funding to renewable energy and energy-efficient technologies (RE & EE). Recent Ex-Im Bank beneficiaries in the solar sector include First Solar, which received a $455.7 million guarantee to support exports of 90 MW of modules to Canada[91] and a $19 million guarantee for exports to India.[92]

U.S. Government Support for Solar Power

Federal policies favoring development of a domestic solar power sector include support for the U.S. solar PV manufacturing industry as well as incentives for solar generation of electricity.

- An advanced energy manufacturing tax credit (MTC) was aimed at supporting renewable energy manufacturers. It reached its funding cap in 2010.
- The Section 1705 Loan Guarantee Program directs funds to manufacturing facilities that employ “new or significantly improved” technologies.
- The investment tax credit (ITC) provides financial incentives for solar power. It is in effect to the end of 2016.
- The Section 1603 Treasury Cash Grant Program requires solar projects to begin construction by December 31, 2011 and be in service by December 31, 2012.

- The Sunshot Initiative is one of several U.S. Department of Energy (DOE) programs to support the solar industry and increase domestic PV manufacturing.

Advanced Energy Manufacturing Tax Credit (MTC)

The Advanced Energy Manufacturing Tax Credit (MTC), Section 48C, which was included in the American Recovery and Reinvestment Act of 2009,[93] provided a 30% tax credit to advanced energy manufacturers that invested in new, expanded, or reequipped manufacturing facilities built in the United States. Solar panel manufacturing was among the 183 projects funded through the MTC before reaching its cap of $2.3 billion in 2010.[94] Solar PV manufacturers benefiting from the credit including Miasole, Calisolar, First Solar, Suniva, Yingli, SunPower, Suntech, and Sharp. Plants receiving the credit have until February 17, 2013 to begin operations. Selected manufacturers of solar PV, and other solar products, that received tax credits under the 48C program are listed in Appendix **Table A-1**. The Obama Administration has requested another $5 billion for the 48C credit. An extension of the MTC has been proposed through the Security in Energy and Manufacturing Act of 2011 (S. 591), or SEAM Act.[95] That bill would make one significant change from the original MTC: higher priority would be given to facilities that manufacture—rather than assemble—goods in the United States.

DOE Loan Guarantee Programs

The Section 1705 loan program, a temporary ARRA program administered by the Department of Energy, provided loan guarantees for renewable energy projects, including solar manufacturing and solar power generation projects. A recent Congressional Research Service report found that 82% of the Section 1705 loan guarantees, or $13.27 billion, have been for solar projects.[96] Specifically, sixteen solar projects, including four manufacturing projects, benefitted from the loan guarantee program before it expired on September 30, 2011 (see **Table 6**).[97] One of the manufacturers, Solyndra, declared bankruptcy in late 2011 and defaulted on its $535 million loan. The other three solar manufacturers are subject to the same market conditions and risks that contributed to the bankruptcy of Solyndra. Recently, Abound Solar announced that it would temporarily eliminate nearly 200 full-time jobs at its manufacturing facility in Colorado.[98]

Table 6. 1705 Loan Guarantees for Solar Generation and Manufacturing Projects

Project	Technology	Loan Guarantee Amount	Location
1366 Technologies	Solar Manufacturing	$150 million	Lexington, MA
Abound Solar	Solar Manufacturing	$400 million	Longmont, CO and Tipton, IN
SoloPower	Solar Manufacturing	$197 million	Portland, OR
Solyndra	Solar Manufacturing	$535 million	Fremont, CA
Abengoa Solar (Mojave Solar)	Solar Generation	$1.2 billion	San Bernardino County, CA
Abengoa Solar (Solana)	Solar Generation	$1.446 billion	Gila Bend, AZ
BrightSource Energy	Solar Generation	$1.6 billion	Baker, CA
Cogentrix of Alamosa	Solar Generation	$90.6 million	Alamosa, CO
Exelon (Antelope Valley Solar Ranch)	Solar Generation	$646 million	Lancaster, CA
Mesquite Solar 1 (Sempra Mesquite)	Solar Generation	$337 million	Maricopa County, AZ
NextEra Energy Resources (Desert Sunlight)	Solar Generation	partial guarantee of $1.46 billion	Riverside County, CA
NextEra Energy Resources (Genesis Solar)	Solar Generation	partial guarantee of $852 million	Riverside County, CA
NRG Energy (California Valley Solar Ranch)	Solar Generation	$1.237 billion	San Luis Obispo, CA
NRG Solar (Agua Caliente)	Solar Generation	$967 million	Yuma County, AZ
Prologis (Project Amp)	Solar Generation	partial guarantee of $1.4 billion	28 States
SolarReserve (Crescent Dunes)	Solar Generation	$737 million	Nye County, NV

Source: U.S. Department of Energy, Loan Guarantee Programs Office, https://lpo.energy.gov

Notes: The 1705 loan guarantee program expired on September 30, 2011.

Recently, the Department of Energy announced that pending applications that were not considered under the 1705 program due to eligibility requirements or time constraints around the September 30, 2011 deadline could be considered for loan guarantees under the Section 1703 loan program,[99] which was part of the Energy Policy Act of 2005.[100] The 1703 program includes loans for renewable energy projects that employ "new or significantly improved" technologies that are not yet in commercial use.[101]

Investment Tax Credit (ITC)

The Investment Tax Credit was first adopted in 2005 as part of the Energy Policy Act of 2005,[102] extended for one additional year in the Tax Relief and Health Care Act of 2006,[103] and again for eight years in the Emergency Economic Stabilization Act of 2008.[104] The ITC, allowing residential and commercial owners of solar projects to offset 30% of a solar system's cost through tax credits, is in place through the end of 2016. In practice, developers of utility-scale solar projects often do not have sufficient income to benefit from the credit, so projects have been developed through structures that transfer the benefit to third-party "tax equity" investors.

The 2008 economic crisis made the ITC less attractive to solar developers as there were fewer tax equity investors that could benefit from the value of the incentives.[105] In 2009, as part of ARRA, the ITC was modified and a new program was adopted which provided a new tax option for solar power developers: a direct cash grant, which may be taken in lieu of the federal business energy investment tax credit that they were otherwise entitled to receive.

1603 Cash Grant Program

The Section 1603 Treasury Grant program expired at the end of 2011. It allowed owners of renewable energy systems to apply for cash grants to cover 30% of the systems' cost, regardless of their tax liability. By the end of March 2012, the 1603 Treasury Program awarded grants to more than 33,000 solar projects totaling $2.1 billion.[106] While an ITC, which reduces overall tax liability, will still be available for solar projects until 2016, it is viewed as a less favorable incentive than the cash grant.

With the expiration, interested parties without the necessary tax liability will again have to rely on tax equity investors to fully monetize the ITC. One outgrowth of this situation is a developing business in third-party ownership of residential and commercial PV systems, with the outside owner installing and maintaining the systems to take advantage of the tax credit; funding comes from investors in securities backed by system leases or from agreements to purchase the power.

SunShot and other Department of Energy Initiatives

The U.S. Department of Energy, which has set a goal for solar energy to provide 14% of domestic electricity by 2030 and 27% by 2050, runs a number of efforts intended to create a stronger domestic PV manufacturing base, under the SunShot Initiative.[107] These include

- the PV incubator program, which began in 2007 and aims to support promising commercial manufacturing processes and products.[108]
- the PV supply chain and cross-cutting technologies project, which provides up to$20.3 million in funds to non-solar companies that may have technologies and practices that could strengthen the domestic PV industry.[109]
- the Advanced Solar Photovoltaic Manufacturing Initiative, with up to $112.5 million in funding over five years, to advance manufacturing techniques to lower the cost of producing PV panels.[110]
- SUNPATH, which stands for Scaling Up Nascent PV At Home and funded at $50million over two years and aims to increase domestic manufacturing bysupporting industrial-scale demonstration projects for PV modules, cells, substrates, or module components.[111]

A separate DOE program to strengthen PV manufacturing is its Advanced Research Project Agency-Energy program, or ARPA-E, which received $275 million in FY 2012. ARPA-E funds transformative energy research that is not being supported by other parts of DOE or the private sector because of technical and financial uncertainty. 1366 Technologies, a silicon PV company, is one solar manufacturer to receive federal funding through this program.[112]

CONCLUSION

Solar manufacturing is currently going through a shakeout, with manufacturers closing U.S. plants because of difficult global business conditions, stiff competition particularly from Chinese companies, and slowing demand for solar panels. Beyond that, the extraction of large quantities of natural gas from shales seems likely to lower the cost of generating electricity from natural gas. While state-level renewable fuels standards, which require utilities to obtain a certain proportion of their electricity from renewable sources, may provide continuing demand for utility-scale PV plants in some states, the lower cost of gas-fired generation may limit interest in large PV installations.

In some parts of the United States, residential and commercial PV systems produce electricity as prices competitive with conventional grid electricity, once subsidies are taken into account. However, although the per-watt cost of solar PV systems has declined significantly, in most areas of the country solar power is still not competitive with conventional grid electricity. The cost disadvantage could widen if subsidies are unavailable or if retail electricity prices decline due to the lower price of natural gas. In the absence of continued government support for solar installations or for the production of solar equipment, the prospects for expansion of domestic PV solar manufacturing may be limited.

APPENDIX.

Table A-1. Solar PV Manufacturers Receiving a 48C Manufacturing Tax Credit
Ranked by size of credit; credits under $1 million excluded

Applicant Name	Tax Credit Requested	Technology Area	Facility State	Updated Descriptions
Hemlock Semi-conductor Corp.	$141,870,000	Solar Components and Materials	MI	To expand polycrystalline plant to capacity of 19,200 metric tons per year
Wacker Polysilicon North America LLC	$128,482,287	Solar Components and Materials	TN	Plant will produce roughly 10 metric tons of pure polysilicon annually
Miasole	$91,350,000	Solar PV	CA	Manufacturing of thin-film solar PV cells and modules
SolarWorld Industries America Inc.	$82,200,000	Solar Components and Materials	OR	To expand its existing 100 MW solar PV manufacturing plant to 500 MW
CaliSolar, Inc.	$51,563,980	Solar CSI	CA	New plant to process silicon feedstock into finished solar cells

Table A-1. (Continued)

Applicant Name	Tax Credit Requested	Technology Area	Facility State	Updated Descriptions
E.I. du Pont de Nemours and Co.	$50,730,000	Solar PV	OH	To expand production of high- performance polyvinyl fluoride films
Nanosolar	$43,453,309	Solar PV	CA	Will make tools for cell manufacture, quality control, and testing
Stion Corporation	$37,500,000	Solar PV	CA	Factory will manufacture high efficiency (11-12%+) CIGS thin-film photovoltaic modules on glass
Xunlight Corporation	$34,500,000	Solar PV	OH	First product is flexible and lightweight thin-film module which can be rolled for shipping
Dow Corning - Solar Silane	$27,300,000	Solar PV	MI	New monosilane facility with 60% of output dedicated to production of amorphous thin-film panels
Jabil Circuit Inc.	$20,400,000	Solar CSI	FL	To retrofit existing plant for PV panel assembly, logistics, procurement, and certification services for cell manufacturers
The Dow Chemical Company	$17,814,621	Solar PV	MI	To produce PV cells built into roofing and siding products
First Solar, Inc.	$16,320,000	Solar PV	OH	Expand plant to produce thin-film modules using cadmium telluride (CdTe) as semiconductor material
Abound Solar, Inc.	$12,600,000	Solar PV	CO	Will expand manufacturing capacity of PV panels using CdTe
Miasole	$10,450,200	Solar PV	CA	Plant will manufacture thin-film solar PV cells and modules
Suniva, Inc.	$5,700,000	Solar CSI	GA	Factory will make monocrystalline silicon-based solar cells
Centrosolar Oregon LLC	$4,740,000	Solar CSI	OR	Plans to build a manufacturing plant for PV solar modules based on crystalline silicon cells
Yingli Green Energy Americas	$4,534,320	Solar CSI	AZ	Plans to open a manufacturing facility to produce PV modules
Solar Power Industries, Inc.	$3,756,000	Solar CSI	PA	Plans to produce multicrystalline cells

Applicant Name	Tax Credit Requested	Technology Area	Facility State	Updated Descriptions
Amonix, Inc	$3,629,998	Solar PV	AZ	To manufacture low-cost solar systems using inexpensive plastic lenses that concentrate sunlight
Sumco Phoenix	$2,674,236	Solar Components and Materials	NM	Plant will manufacture silicon solar blocks
The Dow Chemical Company	$2,220,000	Solar PV	OH	Factory to produce special coatings for use in solar cell manufacture
Suntech	$2,105,848	Solar CSI	AZ	Plans to manufacture poly crystalline solar modules
Spire Semiconductor, LLC	$2,044,500	Solar PV	NH	Will manufacture very high- efficiency concentrator PV cells and receiver assemblies
Solar Power Industries, Inc	$1,611,083	Solar CSI	PA	Plans to produce silicon bricks, wafers, solar power systems, and solar module components
Advanced Energy Industries, Inc.	$1,230,000	Solar Components and Materials	CO	Plans to establish a manufacturing facility for inverters
Applied Photovoltaics, LLC	$1,068,986	Solar PV	NJ	Factory to manufacture solar energy modules for use in building integrated photovoltaics

Source: White House Fact Sheet

Notes: Projects must be commissioned before February 17, 2013.

End Notes

[1] Bloomberg New Energy Finance estimates global module production capacity in 2012 to be 50% in excess of demand; see "Week in Review," vol. 6, issue 131, April 16-23, 2012.

[2] Solar Foundation, *National Solar Jobs Census 2011*, October 2011, p. 25. Its count reflects solar jobs as of August 2011. By comparison, there were 11.7 million jobs in overall U.S. manufacturing in 2011.

[3] U.S. Energy Information Administration (EIA), *Solar Photovoltaic Cell/Module Shipments Report*, January 2012, Table 2, p. 7, http://www.eia.gov/renewable/annual/solar_photo/. Shipments data for 2006 are from Table 3.6 of EIA's 2007 annual PV module/cell manufacturing survey. http://www.eia.gov/renewable/annual/solar_photo/archive/solarpv07.pdf.

[4] SEIA reports that in 2009, cumulative PV installations totaled 99,900. SEIA, *U.S. Solar Market Insight Report*, A4 2011 & 2011 Year-in-Review Full Report, March 2012, pp. 29-30.

[5] SEIA, *U.S. Solar Market Insight Report*, 2011 Year-in-Review Executive Summary, March 2012, p. 3, http://www.slideshare.net/SEIA/us-solar-market-insight-report.

[6] DOE reported that annual installed solar PV capacity grew at a compound annual growth rate of 61.3% between 2000 and 2010, but provided 0.1% of total electricity generation in 2010.

By comparison, U.S. wind installations grew at an compound annual growth rate of 31.6% from 2000 to 2010 and represented 2.3% of total electricity generation in 2010. See pp. 25 and 29 of the U.S. Department of Energy's *2010 Renewable Energy Data Book*, which can be accessed at http://www.nrel.gov/analysis/pdfs/51680.pdf.

[7] The White House, *Blueprint for a Secure Energy Future*, March 30, 2011, p. 32.

[8] Two principal technologies are used in concentrated solar power installations. Concentrating Solar Power (CSP) employs large arrays of mirrors to focus energy on a single point and results in tremendous amounts of heat, creating steam to turn turbines. CSP projects are large-scale and require high initial investment, thus mainly utilities or large tower producers use this technology. Examples of CSP manufacturers include Solargenix, Schott Solar, and Solel. In2010, about 740 MW of CSP was added worldwide, in contrast to the installation of 17 GW of solar PV. See the Duke University report, *Concentrating Solar Power: Clean Energy for the Electric Grid* by Gary Gereffi and Kristen Dubayat http://www.cggc.duke.edu/environment/climatesolutions/greeneconomy_Ch4_ConcentratingSolarPower.pdf.
Concentrated Photovoltaic (CPV) technology, which has been around since the 1970s, uses optics such as lenses to concentrate a large amount of sunlight onto a small area of solar photovoltaic materials to generate electricity. A 2011 report by the National Renewable Energy Laboratory (NREL), *Opportunities and Challenges for Development of a Mature Concentrating Photovoltaic Power Industry*, by Sarah Kurtz, reports that dozens of companies are developing new products for the CPV market, such as Concentrix Solar, Cool Earth Solar, Emcore, Greenvolts, and Energy Innovations. The NREL report can be found at http://www.nrel.gov/docs/fy11osti/43208.pdf.

[9] If the ITC lapses in 2016, businesses will remain eligible for a permanent 10% business tax credit for solar installations and the personal income tax credit for residential installations will end. SEIA, Solar Policies, *The Investment Tax Credit*, http://www.seia.org/cs/solar_policies/solar_investment_tax_credit.

[10] For a detailed discussion of energy tax credits see CRS Report R41953, *Energy Tax Incentives: Measuring Value Across Different Types of Energy Resources*, by Molly F. Sherlock.

[11] ARRA; P.L. 111-5.

[12] CRS Report R41635, *ARRA Section 1603 Grants in Lieu of Tax Credits for Renewable Energy: Overview, Analysis, and Policy Options*, by Phillip Brown and Molly F. Sherlock.

[13] Information about state-level renewable portfolio standards (RPS) can be found on the EIA's website, including an overview of RPS standards, *Most States Have Renewable Portfolio Standards*, January 2012, http://www.eia.gov/todayinenergy/detail.cfm?id=4850.

[14] The Clean Energy Standard Framework announced by the White House in 2011 is discussed in CRS Report R41720, *Clean Energy Standard: Design Elements, State Baseline Compliance and Policy Considerations*, by Phillip Brown.

[15] Feed-in tariffs reimburse renewable energy producers at a set price for the electricity they contribute to the grid. Typical FIT's also have a guaranteed pricing structure for utility companies purchasing the power and often require grid connection. In the United States, FIT policies may require utilities to purchase either electricity, or both electricity and renewable energy attributes from eligible energy generators. A detailed discussion of FIT policy can be founded in the National Renewable Energy Laboratory (NREL) report, *"Feed-In Tariff Policy: Design, Implementation, and RPS Policy Interaction,"* NREL/TP-6A2-45549, March 2009.

[16] Unlike some European countries, Japan continues to support renewable energy. In 2011, it enacted a Renewable Energy Law, which introduced FITs for solar, wind, biomass, geothermal and small hydro effective July 1, 2012.

[17] CleanEdge, *The Texas Solar PV Market: A Competitive Analysis*, 2011, p. 2.

[18] The photovoltaic effect is the basic physical process through which a PV cell converts sunlight into electricity. Sunlight is composed of photons—packets of solar energy. These photons contain different amounts of energy that correspond to the different wavelengths of the solar

spectrum. When photons strike a PV cell, they may be reflected or absorbed, or they may pass right through. The absorbed photons generate electricity.

[19] Fritts made his first cell from selenium. The semiconductor had a thin coat of gold around it and was not very effective in generating electricity. The reason, now known, is that selenium is not a very good semiconductor.

[20] In 1958, PV solar cells received considerable attention because they partially powered the Vanguard 1 satellite launched by the United States. PV cells power nearly all of today's satellites because they can operate for long periods with virtually no maintenance.

[21] Heliotek merged with Spectrolab and produces high-efficiency cells today.

[22] Phech Colatat, Georgeta Vidican, and Richard K. Lester, *Innovation Systems in the Solar Photovoltaic Industry: The Role of Public Research Institutions*, Industrial Performance Center Massachusetts Institute of Technology, Cambridge, MA, June 2009, p. 4, http://web.mit.edu/ipc/research/energy/pdf/EIP_09-007.pdf.

[23] Oil and gas companies used solar power to protect wellheads and underground pipelines from corrosion and to power navigational aids on offshore oil rigs.

[24] Elliott Berman, who founded Solar Power Corporation, pioneered a number of manufacturing changes, including buying cheap solar wafers that had been cast aside by the semiconductor industry, which helped to reduce the cost of solar cells, lowering the selling price from $100 per watt in 1970 to $20 per watt in 1973.

[25] P.L. 9-618. ETA created residential solar credits of up to $2,000 for devices installed on homes. They were in effect from April 20, 1977 to January 1, 1986.

[26] P.L. 95-617. For more information on the history of renewable energy policy see CRS Report RL33588, *Renewable Energy Policy: Tax Credit, Budget, and Regulatory Issues*, by Fred Sissine.

[27] P.L. 109-58

[28] EPAct tax incentives for solar energy applied from January 1, 2006 through December 31, 2007, and the Tax Relief and Health Care Act of 2006 (P.L. 109-432) extended these credits for one additional year. For background on the Solar Investment Tax Credit see SEIA backgrounder, *The Case for the Solar Investment Tax Credit*, SEIA, http://www.seia.org/galleries/pdf/The_Case_for_the_Solar_Investment_Tax_Credit.pdf.

[29] Business Insights, *The Solar Cell Production Global Market Outlook*, June 2011, p. 16. In the 1950s, Bell Labs in New Jersey developed and deployed the first commercial solar cells based on c-Si technology, and Kyocera, a Japanese manufacturer, started mass production in 1983. Today, no U.S.-headquartered manufacturer ranks among the top ten c-Si producers in the world.

[30] Green Rhino Energy, Value Chain Activity: Producing Polysilicon. http://www.greenrhinoenergy.com/solar/industry/ind_01_silicon.php.

[31] Alim Bayaliyev, Julia Kalloz, and Matt Robinson, *China's Solar Policy:*, George Washington University, Subsidies, Manufacturing Overcapacity & Opportunities, December 23, 2011, p. 16, http://solar.gwu.edu/Research/ChinaSolarPolicy_BayaKallozRobins.pdf. The semiconductor industry also uses polysilicon, but increasingly demand for it has shifted to solar PV products.

[32] Two of the world's largest polysilicon manufacturers are U.S.-based companies (Hemlock (a joint venture of Dow Corning and two Japanese manufacturers Shin Etsu and Mitsubishi) and MEMC. European and Japanese manufacturers also rank among the world's leading companies of polysilicon: Renewable Energy Corporation (REC), Wacker-Chemie, Mitsubishi, and Tokuyama. European Photovoltaic Industry Association, *Solar Generation* 6, Solar Photovoltaic Electricity Empowering the World, 2011, p. 27, http://www.greenpeace.org/international/Global/international/publications/climate/2011/Final%20Solar Generation%20VI%20full%20report%20lr.pdf.

[33] U.S. Department of Energy, Energy Efficiency & Renewable Energy, *Solar Photovoltaic Economic Development*, Building and Growing a Local PV Industry, November 2011, p. 53.

[34] European Photovoltaic Industry Association, *Solar Photovoltaic Electricity Empowering the World*, 2011, p. 20.

[35] Green Rhino Energy, Value Chain Activity: Manufacturing Solar Glass, http://www.greenrhinoenergy.com/solar/industry/ind_15_solarglass.php.

[36] AGC Solar, a Belgium-based company that supplies more than half of the world's solar glass, is owned by Asahi Glass of Japan. It produces solar glass for the U.S. market in a factory in Kingsport, TN. Paula Flowers, TN Solar Energy Activities Update, *TN Chamber of Commerce and Industry*, October 7, 2011, p. 6, http://tnchamber.org/environment/2011_F_3_%20Solar%20Update%20by%20Flowers.pdf.

[37] Business Insights, *The Solar Cell Production Global Market Outlook*, June 2011, p. 17. Thin-film cells trace their roots to RCA Laboratories in New Jersey, which fabricated the first a-Si cell in 1976.

[38] Efficiency, which measures the percentage of the sun's energy striking the cell or module, is one important characteristic of a solar cell or module. Over time, average cell efficiencies have increased. EPIA, *Solar Generation 6*, Solar Photovoltaic Electricity Empowering the World, 2011, p. 27.

[39] Several thin-film module manufacturers are facing challenging market conditions. Some announced Chapter 11 bankruptcy in 2010 and 2011, including Solyndra and Energy Conversion Devices, which owns United Solar Ovonic. Miasole, another struggling manufacturer, announced layoffs due to "difficult market conditions."

[40] Bloomberg New Energy Finance, *Joined at the Hip: the U.S.-China Clean Energy Relationship*, May 17, 2010, p. 15, http://www.wilsoncenter.org/sites/default/files/BNEF_joined_at_the_hip_the_us_china_clean_energy_relationship.pdf.

[41] USITC, *Crystalline Silicon Photovoltaic Cells and Modules from China*, Publication 4295, December 2011, p. I-13.

[42] USITC, *Crystalline Silicon Photovoltaic Cells and Modules from China*, Publication 4295, December 2011, pp. 40.

[43] USITC, *Crystalline Silicon Photovoltaic Cells and Modules from China*, Publication 4295, December 2011, pp. V-4.

[44] Alan Goodrich, Ted James, and Michael Woodhouse, *Solar PV Manufacturing Cost Analysis: U.S. Competitiveness in a Global Industry*, National Renewable Energy Laboratory, October 10, 2011, p. 26, http://www.nrel.gov/docs/fy12osti/53938.pdf.

[45] European Photovoltaic Industry Association, *Market Report 2011*, January 2012, p. 4. http://www.epia.org/publications/photovoltaic-publications-global-market-outlook.html.

[46] European Photovoltaic Industry Association, *Market Report 2011*, January 2012, p. 4. http://www.epia.org/publications/photovoltaic-publications-global-market-outlook.html.

[47] SEIA, *U.S. Solar Market Insight Report*, 2011 Year-in-Review Executive Summary, March 2012, p. 3, http://www.slideshare.net/SEIA/us-solar-market-insight-report.

[48] SEIA, *U.S. Solar Market Insight Report*, Q4 2011 & 2011 Year-in-Review Full Report, March 2012, p. 10-17.

[49] Peak gigawatts indicate the amount of power a photovoltaic cell or module will produce at standard test conditions (normally one billion watts per square meter and 25 degrees Celsius).

[50] EIA only began reporting U.S.-manufactured module shipments separately in 2010. In previous years, it reported combined domestically manufactured cell and module shipments, so the data are not directly comparable over time.

[51] Value includes charges for cooperative advertising and warranties, but does not include excise taxes and the cost of freight or transportation. EIA, *Solar Photovoltaic Cell/Module Shipments Report*, January 2012, Table 2, p. 7, http://www.eia.gov/renewable/annual/solar_photo/. Cell shipments totaled nearly $1.2 billion and module shipments reached $5.2 billion.

[52] Data provided to CRS by SEIA based on statistics from its National Solar Database, April 10, 2012.

[53] SolarWorld, with factories in the United States and Europe, is one of the few PV manufacturers with no production facilities in Asia. Production data for SolarWorld are from Photon International's annual cell production survey, *Year of the Tiger*, by Garrett Hering, March 2011, p. 205.

[54] In 2010, Suntech opened its first manufacturing facility in the United States in Goodyear, AZ, with an annual production capacity of 50 MW. Suntech's production capacity in China in that year was 1,800 MW.

[55] SEIA reports 18 PV manufacturing facilities were added in 2009, 22 in 2010, 15 in 2011. These figures do not include manufacturers that may have gone out of business in previous years. The number of new PV facilities is expected to decline to 8 in 2012, 4 in 2013, and 2 in 2014, reports SEIA using information from press reports.

[56] SEIA, *U.S. Solar Market Insight Report*, Q4 2011 & 2011 Year-in-Review Full Report, March 2012, p. 40.

[57] Kate Linebaugh, "GE to Build Solar-Panel Plant in Colorado, Hire 355 People," *Wall Street Journal*, October 13, 2011. http://online.wsj.com/article/ SB10001424052970204002304576629753899008160.html.

[58] Stion, "Stion Announces Grand Opening of New Factory in Mississippi," press release, September 16, 2011, http://www.stion.com/press-releases/110916_Stion_Announces_GrandOpeningofNewFactory.pdf.

[59] Stion, "Stion Announces Commercial Shipments from Hattiesburg, Miss., Factory," press release, March 20, 2012, http://www.stion.com/press-releases/120320_Stion_PVAmerica_HMS.pdf.

[60] The Bureau of Labor Statistics (BLS) does not track employment data for the solar power industry, so the most authoritative data on solar jobs appear to be those in the National Solar Job Census Report, which can be accessed at http://www.solarfoundation.org. The count reported in that census includes jobs not related to PV, such as manufacturing of solar water heating systems.

[61] To address the shortfall in data on the green economy, BLS has undertaken a "green jobs" initiative to measure jobs at establishments that produce green goods and services and use environmentally friendly production processes and practices. Initial data collection efforts are now underway and include the recent release of employment data on green goods and services, see http://www.bls.gov/green.

[62] The Solar Foundation, *National Solar Jobs Census 2011*, October 2011, p. 13. The Solar Foundation collects information on solar industry employment by surveying a "known universe" of firms in various segments of the industry, including construction, manufacturing, and sales and distribution, to fill the gap in government data. The Solar Foundation states that its national job census should be viewed as conservative and there may be more solar workers in the United States than reported in the annual survey.

[63] See, for example, Ben Sills, "Spain Halts Renewable Subsidies to Curb $31 Billion of Debts," *Bloomberg*, January 27, 2012.

[64] EIA, *Solar Photovoltaic Cell/Module Shipments Report 2010*, January 2012, p. 2, http://www.eia.gov/renewable/annual/solar_photo/.

[65] SolarBuzz, "World Solar Photovoltaic Market Grew to 27.4 Gigawatts in 2011, Up 405 Y/Y," press release, March 19, 2012, http://www.solarbuzz.com/our-research/recent-findings/world-solar-photovoltaic-market-grew-274-gigawatts-2011-40-yy.

[66] Christoph Podewils and Beate Knoll, "Crystalline is King," *Photon International*, February 2012, p. 131.

[67] The 2010 PVPS Annual Report shows that exports comprised around 95% of China's production from 2006 to 2010. See Table 9, *PVPS Annual Report 2010*, April 14, 2011, p. 51, http://www.iea-pvps.org/index.php?id=6.

[68] For a comparison of green energy programs and policies in China and the United States, see CRS Report R41748, *China and the United States—A Comparison of Green Energy Programs and Policies*, by Richard J. Campbell.

[69] Coco Liu, "China Uses Feed-in Tariff to Build Domestic Solar Market," *New York Times*, September 14, 2011http://www.nytimes.com/cwire/2011/09/14/14climatewire-china-uses-feed-in-tariff-to-build-domestic-25559.html?pagewanted=1.

[70] European Photovoltaic Industry Association, *Market Report 2011*, January 2012, p. 6, http://www.epia.org.

[71] Government of India, Ministry of New and Renewable Energy, Mission Document, http://www.mnre.gov.in/solar-mission/mission-document-3. Not all of this solar power is expected to come from PV systems.

[72] Arnufl Jager-Waldau, *Research, Solar Cell Production and Market Implementation of Photovoltaics*, European Commission, DG Joint Research Centre, July 2011, p. 83, http://re.jrc.ec.europa.eu/refsys/pdf/PV%20reports/ PV%20Status%20Report%202011.pdf.

[73] The primary harmonized tariff schedule codes covering crystalline silicon PV cells, modules or panels are HTS 8541.40.60.30 (cells) and HTS (8541.40.60.20 (modules), with a few import shipments also falling under HTS 8501.60.00.00 and 8507.20.80.

[74] Andrew David, *U.S. Solar Photovoltaic (PV) Cell and Module Trade Overview*, U.S. International Trade Commission, Executive Briefings on Trade, June 2011, p. 1, http://www.usitc.gov/publications/332/executive_briefings/Solar_Trade_EBOT_Commission_Review_Final2.pdf.

[75] China's largest solar manufacturer, Suntech, was founded in 2001 and went public in 2005. Among the other large Chinese solar manufacturers Trina was founded in 1997, JA Solar in 2005, and Yingli in 2007.

[76] SunPower's solar panels are manufactured at its plant in the Philippines, where it operates six assembly lines with a rated annual solar panel manufacturing capacity of 220 MW. It also uses contract manufacturers in China, Mexico, and Poland to assemble its solar panels. See p. 10 of SunPower's 2010 Annual Report, which can be accessed at http://investors.sunpowercorp.com/annuals.cfm. In 2011, the French oil producer, Total SA, acquired 60% of the company.

[77] Jorge Huacuz Villamar and Jaikme Agredano Diaz, *National Survey Report of PV Power Applications in Mexico*, International Energy Agency, May 2011, p. 10, http://www.iea-pvps.org.

[78] Jane Burgermeister, "South Korea Taps Germany to Help Grow its Solar Industry," *Renewable Energy World*, April 29, 2009. http://www.renewableenergyworld.com/rea/news/article/2009/04/south-korea-looks-to-germany-to-help-grow-its-solar-industry.

[79] Four manufacturers remain anonymous because they fear retaliation by China, possibly with such actions as punitive market access reductions. For more information see the CASM website at http://www.americansolarmanufacturing.org.

[80] In the United States, there are two dispute-resolution systems specifically designed to handle company complaints about apparently anticompetitive trade practices: anti-dumping and countervailing duty mechanisms. The process for antidumping and countervailing duty cases such as the one initiated by CASM can be divided into five stages, each ending with a finding by either the DOC or the ITC. These stages are as follows: 1) initiation of the investigation by the DOC (20 days after filing the petition); 2) the preliminary phase of the ITC's investigation into whether U.S. producers have been injured (with a preliminary determination 45 days after filing of the petition); 3) the preliminary phase of the DOC investigation (with a preliminary determination 115 days after the ITC's determination for antidumping cases or 40 days for countervailing duty cases); 4) the final phase of the DOC investigation (with a final determination 75 days after the DOC's determination) and 5) the final phase of the ITC's investigation.

[81] The DOC preliminarily assessed duties of 2.9% on Suntech, 4.73% for Trina Solar, and 3.61% for all other Chinese producers, which will apply retroactively 90 days. The DOC will make a final determination on its countervailing duty investigation on June 4, 2012. The ITC will rule on the case on July 19, 2012. For a DOC fact sheet, see "Fact Sheet: Commerce Preliminarily Finds Countervailable Subsidization of Crystalline Silicon Photovoltaic Cells,

Whether or Not Assembled into Modules from the People's Republic of China," press release, March 2012, http://ia.ita.doc.gov/download/factsheets/factsheet-prc-solar-cells-adcvd-prelim-20120320.pdf.

[82] CASE claims to represent 150 solar installation firms, retailers, and system owners, and solar panel manufacturers owned or operating in the United States. For additional background, see http://coalition4affordablesolar.org/.

[83] The source of the 50,000 net jobs figure is a CASE commissioned study by the Brattle Group. See Mark Berkman, Lisa Cameron, and Judy Chang, *The Employment Impacts of Proposed Tariffs on Chinese Manufactured Photovoltaic Cells and Modules*, The Brattle Group, January 30, 2012, pp. ES-2-6, http://coalition4affordablesolar.org/wp-content/uploads/2012/01/TBG_Solar-Trade-Impact-Report.pdf.

[84] See the GTM Research studies prepared for SEIA, *U.S. Solar Energy Trade Assessment 2011, Trade Flows and Domestic content for Solar-Related Goods and Services in the United States*, August 2011, pp. 25 and 30 and the November 2010 edition, pp. 25 and 29.

[85] SEIA, *U.S. Solar Energy Trade Assessment 2011*, August 2011, p. 45, see Figure 2-24.

[86] Andrew David and Mihir Torsekar, "An Inside Look at U.S. Solar Imports, Exports," *Solar Industry*, November 2011.

[87] Generally, solar cells and modules enter foreign markets under the harmonized tariff schedule (HTS) 8541.40.60.20 and 8541.40.60.30, which are included in the ITA. The EU, Canada, Japan, India, Malaysia, and China are among its signatories. Missing from the list of ITA members are countries such as Brazil, Mexico, Chile, and South Africa. Background on the ITA can be found on the World Trade Organization website at http://www.wto.org/english/tratop_E/inftec_e/inftec_e.htm.

[88] Silicon enters foreign markets under HTS 2804.61. The EU's applied tariff is zero, China's is 4%, Malaysia's is zero, and the Philippines' is 3%. South Korea's applied tariff is 3% for non-FTA member countries, but because of the U.S.-Korea Free Trade Agreement the duty rate for silicon exports from the United States to South Korea is zero.

[89] Jacob Funk Kirkegaard, Thilo Hanemann, and Lutz Weischer, et al., *Toward a Sunny Future? Global Integration in the Solar PV Industry*, Peterson Institute for International Economics, May 2010, pp. 32-34.

[90] For more information on solar PV policies by country see, Arnulf Jager_Waldau, *PV Status Report 201*, European Commission, July 2011, http://re.jrc.ec.europa.eu/refsys/pdf/PV%20reports/PV%20Status%20Report%202011.pdf.

[91] Export-Import Bank of the United States, "Ex-Im Bank Announces over $455 Million in Project Financing for First Solar's Exports to Canada," press release, September 2, 2011, http://www.exim.gov/pressrelease_print.cfm/830B629B-023E-5C34-5863BEEA2A634632/.

[92] Export-Import Bank of the United States, "Ex-Im Bank Supports Renewable Energy Jobs by Financing Solar Power Projects in India," press release, March 30, 2011, http://www.exim.gov/pressrelease_print.cfm/0C34ED47-DA59-908E-85498C3C62B91BB2/.

[93] The credit was authorized in Section 1302 of the American Recovery and Reinvestment Act.

[94] White House, "President Obama Awards $2.3 Billion for New Clean-Tech Manufacturing Jobs," press release, January 8, 2010, http://www.whitehouse.gov/the-press-office/president-obama-awards-23-billion-new-clean-tech-manufacturing-jobs.

[95] Senator Sherrod Brown, "Sen. Brown Introduces Legislation to Expand Manufacturing Tax Credit," press release, May 6, 2010, http://www.brown.senate.gov/newsroom/press_releases/release/?id=125b64dc-3005-4b71-a6ad-0b96c24a3c73.

[96] The remaining 18% support a variety of projects in other renewable energy sectors, including biofuels, energy storage, wind generation, transmission, and geothermal electricity. See CRS Report R42059, *Solar Projects: DOE Section 1705 Loan Guarantees*, by Phillip Brown.

[97] In April 2012, the Department of Energy announced that it expects to issue conditional loan guarantees "over the next several months" for pending renewable energy projects, including

solar projects. April 5, 2012 letter from David Frantz, Acting Executive Director, Loans Program Office, DOE, http://energy.gov/articles/update-1703-loan-program

[98] See Abound Solar Production Plan FAQ at http://www.abound.com/feb28faq.

[99] An update on the 1703 loan program was announced on April 5, 2012, http://energy.gov/articles/update-1703-loan-program.

[100] P.L. 109-58

[101] 1703 program eligibility is described on DOE's Loan Programs Office website at https://lpo.energy.gov/?page_id=31.

[102] P.L. 109-58

[103] P.L. 109-432

[104] P.L. 110-343

[105] SEIA reported in 2007 there were 20 tax equity providers, which dropped to only 11 in 2009. For additional background see SEIA, *The Crisis in the Tax Equity Market and the Need to Extend the Treasury Grant Program*, September 2010, p. 3, http://seia.org/galleries/pdf/Tax_Equity_Crisis_Slides.pdf.

[106] U.S. Department of Treasury, *Overview and Status Update of the 1603 Program*, March 29, 2012, p. 2, http://www.treasury.gov/initiatives/recovery/Documents/Status%20overview.pdf.

[107] U.S. Department of Energy, *SunShot Vision Study*, February 2012, http://www1.eere.energy.gov/solar/pdfs/47927.pdf.

[108] U.S. Department of Energy, Photovoltaic Technology Incubator, http://www1.eere.energy.gov/solar/pv_incubator.html.

[109] U.S. Department of Energy, *Photovoltaic Supply Chain and Cross-Cutting Technologies*, http://www1.eere.energy.gov/solar/sunshot/pv_supply_chain.html.

[110] For more information on these programs, see Department of Energy, SunShot Photovoltaic Manufacturing Initiative, http://www1.eere.energy.gov/solar/sunshot/pvmi.html.

[111] EERE, Funding Opportunity Announcements, SUNPATH Part 2, https://eere-exchange.energy.gov/#521ef7df-e162-4db5-9cf1-7dafd431307f.

[112] 1366 Technologies, "1336 Technologies Awarded Four Million in ARPA-E Funding," press release, October 26, 2009, http://www.1366tech.com/1366-technologies-awarded-four-million-in-arpa-e-funding/.

In: Solar Photovoltaics
Editors: J. E. Foster and A. M. Powell
ISBN: 978-1-62257-234-2

Chapter 2

SOLAR PHOTOVOLTAIC CELL/ MODULE SHIPMENTS REPORT 2010*

U.S. Department of Energy

OVERVIEW

Despite a sluggish U.S. economy in 2010, the U.S. photovoltaic (PV) industry saw strong growth. Total shipments of PV modules in 2010 more than doubled compared to total module shipments in 2009, rising from nearly 1.2 peak gigawatts (GWdc[1] peak) to more than 2.6 peak gigawatts (Figure 1). This surge in growth was supported in part by rapid declines in the price of PV cells and modules (Figure 2), and by government incentives and policies at the Federal, state and local levels.

As a consequence of this strong growth, favorable business opportunities emerged for companies to expand or enter into the PV market. In 2010, there were 112 active PV manufacturers, importers and exporters that shipped PV cells and modules (Table 1). This marks an increase of nearly 11 percent from the 101 companies doing business in 2009.

* This is an edited, reformatted and augmented version of the Independent Statistics & Analysis, U.S. Department of Energy, dated January 2012.

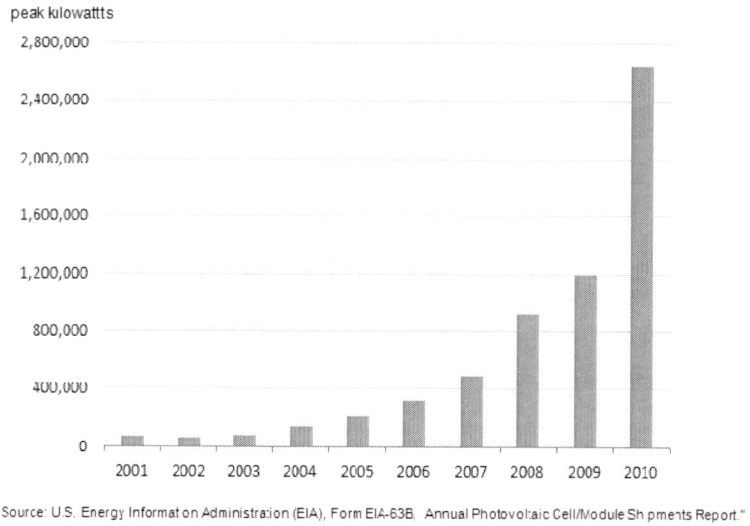

Figure 1. Annual photovoltaic module shipments, 2001-2010.

INDUSTRY STATUS

The PV module market continues to be a very dynamic and competitive market. Many of the companies reporting PV shipments in 2010 also reported being involved in one or more of the following photovoltaic-related activities (Table 1):

- 57 companies were involved in module and/or cell manufacturing.
- 60 designed modules or systems.
- 38 developed prototype modules.
- 21 developed prototype systems.
- 61 were involved in wholesale distribution.
- 26 were involved in retail distribution.
- 29 installed PV systems.

In addition, several manufacturers are planning to introduce new photovoltaic-related technical products in the next calendar year (Table 1):

- 24 plan to introduce new single-crystal silicon modules.
- 14 plan to introduce new cast silicon modules.

- 13 plan to introduce new thin-film modules.
- 8 plan to introduce new concentrator photovoltaic modules.

Employment in PV-related activities increased more than 21 percent, from 14,443 full-time equivalent (FTE)[2] employees in 2009 to 17,487 FTE in 2010. Of the 112 companies, 83 had 90 percent or more of their companywide revenues in PV-related activities, 8 had 50 to 89 percent, 6 had 10 to 49 percent, and 15 had less than 10 percent (Table 1).

Table 1. U.S. photovoltaic industry status, 2010

Reporting Companies, total number of:	112
Employment, full-time equivalent employees:	17,487
Involved in Photovoltaic-related Activities, number of reporting companies:	
Module and/or Cell Manufacturing	57
Module or System Design	60
Prototype Module Development	38
Prototype Systems Development	21
Wholesale Distribution	61
Retail Distribution	26
Installation	29
Non-module System Component	9
Sales as a Percent of Total Company Revenue, number of reporting companies:	
90-100%	83
50-89%	8
10-49%	6
Less than 10%	15
Companies Expecting to Introduce New Photovoltaic-related Products in 2011, by product type and number:	
Crystalline Silicon	
Single-Crystal Silicon Modules	24
Cast Silicon Modules	14
Ribbon Silicon Modules	-
Thin-Film	
Amorphous Silicon Modules	7
Other (Thin-Film Modules)	6
Concentrators	8
Non-module Components	6

\- = No data reported

Source: U.S. Energy Information Administration, Form EIA-63B, "Annual Photovoltaic Cell/Module Shipments Report."

VALUE, AVERAGE PRICE, AND AVERAGE EFFICIENCY

In 2010, the value of photovoltaic cell shipments totaled nearly $1.2 billion and the value of photovoltaic module shipments reached nearly $5.2 billon (Table 2). Value includes charges for cooperative advertising and warranties, but does not include excise taxes and the cost of freight or transportation.

Table 2. Value, average price and average efficiency of photovoltaic cell and module shipments by type, 2010

Value, total shipments (cells):	thousand dollars
Total Cells:	$1,173,355
Crystalline Silicon	$1,136,827
Thin-Film	$17,548
Concentrator	$18,980
Price, average value (cells):	dollars per watt
Total Cells:	$1.13
Crystalline Silicon	$1.15
Thin-Film	$1.35
Concentrator	$0.48
Value, total shipments (modules):	thousand dollars
Total Modules:	$5,192,669
Crystalline Silicon	$4,290,698
Thin-Film	$875,127
Concentrator	$26,844
Price, average value (modules):	dollars per watt
Total Modules:	$1.96
Crystalline Silicon	$2.03
Thin-Film	$1.68
Concentrator	$2.66
Average Energy Conversion Efficiency, modules shipped:	percent per peak kilowatt
Crystalline Silicon	16%
Thin-Film	11%
Concentrator	27%

Source: U.S. Energy Information Administration, Form EIA-63B, "Annual Photovoltaic Cell/Module Shipments Report."

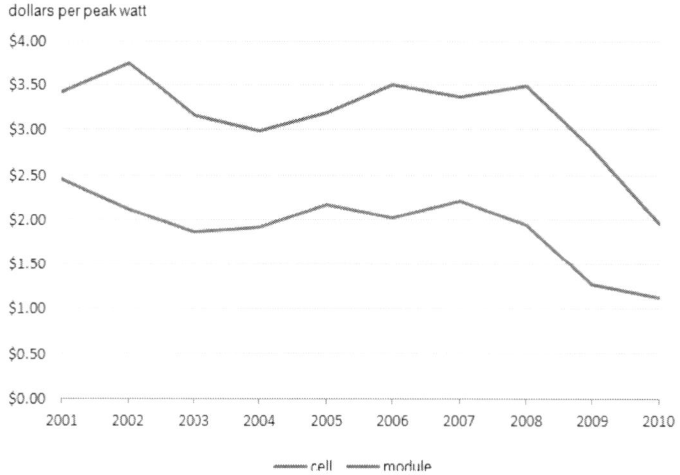

Source: U.S. Energy Information Administration (EIA), Form EIA-63B, "Annual Photovoltaic Cell/Module Shipments Report."

Figure 2. Average price of photovoltaic cells and modules, 2001-2010.

Average prices of photovoltaic cells and /modules per peak watt of capacity shipped have declined steady over the past three years due, in part, to government incentives and market competition. For photovoltaic cells, the average price has decreased almost 11 percent, from $1.27 in 2009 to $1.13 in 2010, and the average price of photovoltaic modules fell nearly 30 percent, from $2.79 in 2009 to $1.96 in 2010 (Figure 2).

The performance of a photovoltaic module is typically measured in terms of its energy conversion efficiency, i.e., the percentage of incident solar energy (input) that the module converts to electricity (output) under standard rating conditions. In 2010, the average energy conversion efficiencies were as follows (Table 2):

- 16 percent for Crystalline Silicon PV module.
- 11 percent for Thin-film PV module.
- 27 percent for Concentrator PV module.

U.S. Manufactured

In 2010, 1,066,867 peak kilowatts of PV modules were manufactured in the United States. Manufacturers in California, Massachusetts, Michigan,

Ohio, and Tennessee produced more than 82 percent of the domestically manufactured PV modules (Table 5). Crystalline silicon accounted for more than 59 percent of the annual domestically manufactured PV modules, followed by thin-film modules with nearly 40 percent. Together, crystalline silicon and thin-film accounted for more than 99 percent of all PV manufacturing.

Table 3. Source and disposition of photovoltaic cell shipments by type, 2010

peak kilowatts				
Cell Shipments	**Type**			
	Crystalline Silicon	**Thin-Film**	**Concentrator**	**Total**
Source				
Inventory, Start-of-Year	45,852	2,065	6,040	53,957
Manufactured during Reporting Year	508,904	11,794	33,267	553,964
Imported during Reporting Year	466,228	48	2,225	468,501
Purchased from U.S. Original Equipment Manufacturer	16,092	-	7,134	23,226
Total Available For Shipment	1,037,076	13,907	48,666	1,099,649
Disposition				
Cells Assembled into Modules	634,975	536	6,775	642,286
Sales to U.S. Original Equipment Manufacturer for Resale	206,690	288	24,456	231,434
Export Shipments	144,455	12,179	8,385	165,019
Total Shipments	986,120	13,003	39,616	1,038,739
Inventory, End-of-Year	50,956	904	9,050	60,910

- = No data reported.

Note: Totals may not equal sum of components due to independent rounding.

Source: U.S. Energy Information Administration, Form EIA-63B, "Annual Photovoltaic Cell/Module Shipments Report."

Table 4. Source and disposition of photovoltaic module shipments by type, 2010

peak kilowatts				
Module Shipments	**Type**			
	Crystalline Silicon	**Thin-Film**	**Concentrator**	**Total**
Source				
Inventory, Start-of-Year	118,645	78,057	5,388	202,090
Manufactured during Reporting Year	633,730	426,260	6,878	1,066,867
Imported during Reporting Year	1,622,315	105,152	6,682	1,734,149
Purchased from U.S. Original Equipment Manufacturer	21,546	1,035	-	22,581
Total Available For Shipment	2,396,235	610,504	18,947	3,025,686
Disposition				
U.S. Shipments	1,404,873	255,165	7,505	1,667,543
Export Shipments	710,007	264,351	2,596	976,955
Total Shipments	2,114,881	519,516	10,101	2,644,498
Inventory, End-of-Year	281,354	90,988	8,846	381,189

- = No data reported.

Note: Totals may not equal sum of components due to independent rounding.

Source: U.S. Energy Information Administration, Form EIA-63B, "Annual Photovoltaic Cell/Module Shipments Report."

IMPORTS

Imports of PV modules shipments totaled 1,734,149 peak kilowatts in 2010. The predominant type of import shipment was crystalline silicon modules, accounting for more than 93.5 percent (1,622,315 peak kilowatts) of total imports. China, Philippines, and the Mexico accounted for about than 88 percent of total imports (Table 6), respectively 46 percent, 31 percent, and 11 percent.

Table 5. Origin of U.S. manufactured photovoltaic module shipments by State and type, 2010

peak kilowatts					
State	Type			Total	Percent of U.S. Total
	Crystalline Silicon	Thin-Film	Concentrator		
Arizona	25,321	436	-	25,757	2.41%
California	167,587	88,881	4,514	260,982	24.46%
Colorado	-	30,287	-	30,287	2.84%
Delaware	22,172	-	-	22,172	2.08%
Florida	3	-	-	3	0.00%
Illinois	1,593	-	-	1,593	0.15%
Iowa	-	1,460	-	1,460	0.14%
Massachusetts	157,088	45	-	157,133	14.73%
Michigan	-	97,639	-	97,639	9.15%
New Mexico	70,862	-	2,358	73,220	6.86%
New York	1,196	-	6	1,202	0.11%
Ohio	-	207,512	-	207,512	19.45%
Oregon	33,105	-	-	33,105	3.10%
Tennessee	152,526	-	-	152,526	14.30%
Texas	1,650	-	-	1,650	0.15%
Washington	626	-	-	626	0.06%
Total	633,730	426,260	6,878	1,066,867	100.00%

- = No data reported.

Note: Totals may not equal sum of components due to independent rounding.

Source: U.S. Energy Information Administration, Form EIA-63B, "Annual Photovoltaic Cell/Module Shipments Report."

EXPORTS

Exports of PV module shipments totaled 976,955 peak kilowatts in 2010, accounting for about 37 percent of total shipments, or approximately 56 percent of PV module imports. The predominant type of export shipment was crystalline silicon modules, accounting for nearly 73 percent (710,007 peak kilowatts) of total exports. The export market was dominated by sales to Germany, Italy, and Canada (Table 7), respectively 34 percent, 27 percent, and 11 percent.

Table 6. Origin of photovoltaic module import shipments by country and type, 2010

peak kilowatts

Country	Type			Total	Percent of U.S. Total
	Crystalline Silicon	Thin-Film	Concentrator		
Argentina	290	-	-	290	0.02%
Canada	143	-	-	143	0.01%
China	792,233	1,069	5,663	798,965	46.07%
France	197	-	-	197	0.01%
Germany	982	-	1,000	1,982	0.11%
Hong Kong	1,525	-	-	1,525	0.09%
India	66,010	-	-	66,010	3.81%
Japan	6,913	53,505	-	60,417	3.48%
Korea, South	1,906	-	-	1,906	0.11%
Mexico	141,341	50,579	19	191,939	11.07%
Norway	2,039	-	-	2,039	0.12%
Philippines	534,479	-	-	534,479	30.82%
Singapore	33,000	-	-	33,000	1.90%
Spain	554	-	-	554	0.03%
Sweden	11,000	-	-	11,000	0.63%
Taiwan	29,682	-	-	29,682	1.71%
Thailand	22	-	-	22	0.00%
Total	1,622,315	105,152	6,682	1,734,149	100.00%

- = No data reported.

Note: Totals may not equal sum of components due to independent rounding.

Source: U.S. Energy Information Administration, Form EIA-63B, "Annual Photovoltaic Cell/Module Shipments Report."

U.S. SHIPMENTS

Shipment by Geography

U.S. PV module shipments totaled 1,667,543 peak kilowatts in 2010. The shipments went to all 50 States, the District of Columbia, American Samoa, Guam, Puerto Rico, and the Virgin Islands (Table 8).

Table 7. Destination of photovoltaic module export shipments by country and type, 2010

peak kilowatts					
Country	Type			Total	Percent of U.S. Total
	Crystalline Silicon	Thin-Film	Concentrator		
Argentina	179	-	-	179	0.02%
Australia	19,260	548	227	20,034	2.05%
Austria	1,161	420	-	1,581	0.16%
Belgium	7,880	8,398	-	16,278	1.67%
Bermuda		-	-	1	*
Brazil	1,594	34	-	1,628	0.17%
Bulgaria	-	51	-	51	0.01%
Canada	55,782	48,924	4	104,710	10.72%
Chile	265	51	-	316	0.03%
China	1,977	1,874	908	4,760	0.49%
Colombia	32	-	-	32	*
Costa Rica	260	-	-	260	0.03%
Czech Republic	19,739	1,942	-	21,681	2.22%
Denmark	176	18	-	194	0.02%
Dominican Republic	7	-	-	7	*
Finland	-	6	-	6	*
France	49,332	28,026	s	77,359	7.92%
French Polynesia	706	-	-	706	0.07%
Georgia	-	2	-	2	*
Germany	250,852	85,078	2	335,931	34.39%
Greece	5,334	2,314	1,252	8,900	0.91%
Haiti	230	-	-	230	0.02%
Honduras	-	2	-	2	*
Hong Kong	-	295	-	295	0.03%
Hungary	-	1	-	1	*
India	1,953	2,075	-	4,028	0.41%
Indonesia	-	1	-	1	*
Iraq	100	-	-	100	0.01%
Ireland	-	14	-	14	*
Israel	2,556	209	-	2,765	0.28%
Italy	194,270	66,163	-	260,433	26.66%
Jamaica	119	-	-	119	0.01%
Japan	17,237	7,921	-	25,158	2.58%
Kenya	315	-	-	315	0.03%
Korea, South	6,873	1,470	17	8,360	0.86%
Latvia	-	1	-	1	*
Lebanon	216	-	-	216	0.02%

Country	Type			Total	Percent of U.S. Total
	Crystalline Silicon	Thin-Film	Concentrator		
Luxembourg	3,165	-	-	3,165	0.32%
Macedonia (Skopje)	100	-	-	100	0.01%
Malaysia	1	1	8	10	*
Malta	118	-	8	126	0.01%
Mexico	104	123	25	253	0.03%
Netherlands	1,365	755	-	2,120	0.22%
New Zealand	-	90	-	90	0.01%
Nicaragua	4	-	-	4	*
Nigeria	547	-	-	547	0.06%
Norway	-	5	-	5	*
Peru	163	9	-	172	0.02%
Philippines	125	-	-	125	0.01%
Portugal	440	229	10	679	0.07%
Russia	-	130	-	130	0.01%
Saudi Arabia	1	1,365	126	1,492	0.15%
Sierra Leone	53	-	-	53	0.01%
Singapore	11,543	2	-	11,545	1.18%
Slovenia	1,216	7	-	1,223	0.13%
South Africa	60	200	8	268	0.03%
Spain	42,318	5,257	-	47,575	4.87%
Sweden	-	3	-	3	*
Switzerland	6,353	226	-	6,579	0.67%
Taiwan	8	26	-	34	*
Tajikistan	250	-	-	250	0.03%
Thailand	187	25	-	212	0.02%
Trinidad and Tobago	1	-	-	1	*
Turkey	-	7	-	7	*
United Arab Emirates	1,465	-	-	1,465	0.15%
United Kingdom	2,013	56	-	2,069	0.21%
Venezuela	1	-	-	1	*
Total	710,007	264,351	2,596	976,955	100.00%

* = Less than 0.01 percent.

s = Value is less than 0.5 of the table metric; value is included in any associated total.

- = No data reported.

Note: Totals may not equal sum of components due to independent rounding.

Source: U.S. Energy Information Administration, Form EIA-63B, "Annual Photovoltaic Cell/Module Shipments Report."

Nearly 70 percent of U.S. PV module shipments (1,163,992 peak kilowatts) went to five States (in order of decreasing volume): California, New

Jersey, Arizona, Nevada, and Colorado. More than 54 percent (903,705 peak kilowatts) of domestic shipments went to California (39 percent) and New Jersey (15 percent).

Table 8. U.S. photovoltaic module shipments by State/territory and type, 2010

peak kilowatts					
State/Territory	Type			Total	Percent of U.S. Total
	Crystalline Silicon	Thin-Film	Concentrator		
Alabama	242	s	-	243	0.01%
Alaska	99	-	-	99	0.01%
American Samoa	677	-	-	677	0.04%
Arizona	106,855	3,474	2,312	112,641	6.75%
Arkansas	104	475		579	0.03%
California	598,030	56,042	3,112	657,184	39.41%
Colorado	66,315	694	1,005	68,014	4.08%
Connecticut	8,205	1,150	-	9,355	0.56%
Delaware	8,824	1,141	-	9,965	0.60%
District of Columbia	390	-	-	390	0.02%
Florida	25,217	11,979	-	37,196	2.23%
Georgia	5,349	348	-	5,697	0.34%
Guam	7	-	-	7	*
Hawaii	17,102	707	-	17,809	1.07%
Idaho	250	35	-	284	0.02%
Illinois	3,585	14	-	3,599	0.22%
Indiana	2,356	1,178	-	3,534	0.21%
Iowa	109	4	-	113	0.01%
Kansas	496	62	-	558	0.03%
Kentucky	682	141	-	823	0.05%
Louisiana	2,509	31	-	2,540	0.15%
Maine	614	4	-	618	0.04%
Maryland	26,823	5,451	-	32,274	1.94%
Massachusetts	32,696	301	-	32,997	1.98%
Michigan	6,371	1,018	-	7,389	0.44%
Minnesota	1,813	49	-	1,862	0.11%
Mississippi	1	s	-	1	*
Missouri	969	86	-	1,056	0.06%
Montana	575	10	-	585	0.04%
Montana	575	10	-	585	0.04%
Nebraska	171	41	-	212	0.01%
Nevada	16,470	63,162	-	79,632	4.78%

State/Territory	Type			Total	Percent of U.S. Total
	Crystalline Silicon	Thin-Film	Concentrator		
New Hampshire	4,450	41	-	4,491	0.27%
New Jersey	231,546	14,974	-	246,521	14.78%
New Mexico	21,818	37,790	1,000	60,608	3.63%
New York	46,118	9,320	-	55,437	3.32%
North Carolina	29,172	22	-	29,194	1.75%
North Dakota	116	-	-	116	0.01%
Ohio	6,075	17,026	-	23,101	1.39%
Oklahoma	679	60	-	739	0.04%
Oregon	7,474	92	-	7,567	0.45%
Pennsylvania	44,636	7,568	-	52,204	3.13%
Puerto Rico	8,550	22	-	8,572	0.51%
Rhode Island	1,947	1	-	1,948	0.12%
South Carolina	451	1	-	452	0.03%
South Dakota	102	1	-	103	0.01%
Tennessee	5,372	52	-	5,424	0.33%
Texas	39,876	18,683	77	58,636	3.52%
Utah	1,505	10	-	1,515	0.09%
Vermont	10,388	8	-	10,396	0.62%
Virgin Islands of the U.S.	83	-	-	83	*
Virginia	2,541	77	-	2,618	0.16%
Washington	4,607	90	-	4,698	0.28%
West Virginia	426	2	-	428	0.03%
Wisconsin	2,145	1,722	-	3,867	0.23%
Wyoming	888	6	-	894	0.05%
Total	1,404,873	255,165	7,505	1,667,543	100.00%

* = Less than 0.01 percent. s = Value is less than 0.5 of the table metric; value is included in any associated total. - = No data reported.

Note: Totals may not equal sum of components due to independent rounding.

Source: U.S. Energy Information Administration, Form EIA-63B, "Annual Photovoltaic Cell/Module Shipments Report."

Shipments by Sector

U.S. PV module shipments to the commercial sector in 2010 accounted for 908,224 peak kilowatts, slightly more than 54 percent of the domestic market. Of the domestic shipments to the commercial sector, approximately 94 percent was crystalline silicon, about 6.2 percent was thin-film PV, and less than 0.3 percent was concentrator PV (Table 9).

Table 9. U.S. photovoltaic module shipments by sector, end use, and type, 2010

peak kilowatts					
Sector	End Use	Type: Crystalline Silicon	Type: Thin-Film	Type: Concentrato r	Total
Residential	Grid-connected Centralized PV System	-	-	-	-
	Grid-connected Distributed PV System	427,041	35,376	-	462,417
	Off-grid Domestic PV System	4,489	221	-	4,710
	Off-grid Non-domestic PV System	31	6	-	37
Total		431,562	35,603	-	467,165
Commercial	Grid-connected Centralized PV System	-	-	-	-
	Grid-connected Distributed PV System	847,642	55,473	2,192	905,306
	Off-grid Domestic PV System	13	-	-	13
	Off-grid Non-domestic PV System	1,871	1,034	-	2,905
Total		849,526	56,506	2,192	908,224
Industrial	Grid-connected Centralized PV System	-	-	-	-
	Grid-connected Distributed PV System	39,738	-	1,001	40,739
	Off-grid Domestic PV System	6,218	-	-	6,218
	Off-grid Non-domestic PV System	1,185	66	-	1,251
Total		47,141	66	1,001	48,208
Electric Power	Grid-connected Centralized PV System	76,645	162,989	4,313	243,947
	Grid-connected Distributed PV System	-	-	-	-
	Off-grid Domestic PV System	-	-	-	-
	Off-grid Non-domestic PV System	-	-	-	-
Total		76,645	162,989	4,313	243,947
Total U.S. Photovoltaic Module Shipments		1,404,873	255,165	7,505	1,667,543

- = No data reported.

Note: Totals may not equal sum of components due to independent rounding.

Source: U.S. Energy Information Administration, Form EIA-63B, "Annual Photovoltaic Cell/Module Shipments Report."

The residential sector was the second-largest domestic market in 2010, accounting for 467,165 peak kilowatts, or 28 percent of the domestic market share. About 92 percent of its shipments was crystalline silicon and more than 7 percent was thin-film.

Shipments to the electric power sector amounted to 243,947 peak kilowatts, or nearly 15 percent of the domestic market share. Crystalline silicon accounted for more than 31 percent of the electric power shipments, thin-film accounted for nearly 67 percent, and concentrator accounted for 0.2 percent.

The industrial sector, with less than 3 percent of domestic shipments, was the smallest domestic sales market, totaling 48,208 peak kilowatts. Nearly 98 percent of the shipments was crystalline silicon, about 0.1 percent was thin-film, and just over 2 percent was concentrator.

Shipments by Grid or Off-Grid

U.S. PV module shipments to grid-connected distributed PV systems in 2010 accounted for 1,408,462 peak kilowatts, or more than 84 percent of the domestic market. Of the domestic shipments to grid-connected distributed PV systems, more than 93 percent was crystalline silicon, more than 6 percent was thin-film, and less than 0.3 percent was concentrator (Table 10).

U.S. PV module shipments to grid-connected centralized PV systems were the second-largest domestic market in 2010, accounting for 243,947 peak kilowatts, or close to 15 percent of the domestic market share. More than 31 percent of shipments was crystalline silicon, nearly 67 percent was thin-film, and less than 2 percent was concentrator.

PV module shipments to off-grid PV systems such as households and communities not connected to the utility grid amounted to 10,941 peak kilowatts, or less than 1 percent of the domestic market share. About 98 percent of total domestic shipments was crystalline silicon and 2 percent was thin-film.

Other off-grid PV systems such as water pumping, remote communications, safety and protection devices, at locations without the presence of the utility grid, accounted for less than 0.3 percent of domestic shipments. This was the smallest domestic end-use market, totaling 4,193 peak kilowatts. Nearly 74 percent was crystalline silicon and about 26 percent was thin-film.

Table 10. U.S. photovoltaic module shipments by end use, sector, and type, 2010

peak kilowatts					
End Use	**Sector**	**Type**			**Total**
		Crystalline Silicon	**Thin-Film**	**Concentrator**	
Grid-connected Centralized PV System	Residential	-	-	-	-
	Commercial	-	-	-	-
	Industrial	-	-	-	-
	Electric Power	76,645	162,989	4,313	243,947
Total		76,645	162,989	4,313	243,947
Grid-connected Distributed PV System	Residential	427,041	35,376	-	462,417
	Commercial	847,642	55,473	2,192	905,306
	Industrial	39,738	-	1,001	40,739
	Electric Power	-	-	-	-
Total		1,314,422	90,848	3,193	1,408,462
Off-grid Domestic PV System	Residential	4,489	221	-	4,710
	Commercial	13	-	-	13
	Industrial	6,218	-	-	6,218
	Electric Power	-	-	-	-
Total		10,719	221	-	10,941
Off-grid Non-domestic PV System	Residential	31	6	-	37
	Commercial	1,871	1,034	-	2,905
	Industrial	1,185	66	-	1,251
	Electric Power	-	-	-	
Total		3,088	1,106	-	4,193
Total U.S. Photovoltaic Module Shipments		1,404,873	255,165	7,505	1,667,543

- = No data reported.

Note: Totals may not equal sum of components due to independent rounding.

Source: U.S. Energy Information Administration, Form EIA-63B, "Annual Photovoltaic Cell/Module Shipments Report."

INVENTORY

In 2010, U.S. inventories of PV module at yearend were 381,189 peak kilowatts. Ending inventories of U.S. PV modules in 2010 were almost 89 percent or 179,099 peak kilowatts higher than the year-end inventories of the previous year, which totaled 202,090 peak kilowatts (Table 4). The significant year-over-year increase in inventory level was in line with the growth in production and shipments.

Table 11. U.S. photovoltaic module shipments to residential sector by State/territory and type, 2010

peak kilowatts					
State/Territory	Type			Total	Percent of U.S. Total
	Crystalline Silicon	Thin-Film	Concentrator		
Alabama	157	s	-	158	0.03%
Alaska	69	-	-	69	0.01%
Arizona	44,329	28	-	44,357	9.50%
Arkansas	104	s	-	104	0.02%
California	208,612	16,726	-	225,338	48.24%
Colorado	13,905	232	-	14,137	3.03%
Connecticut	2,509	5	-	2,514	0.54%
Delaware	4,373	1	-	4,374	0.94%
District of Columbia	251	-	-	251	0.05%
Florida	6,088	67	-	6,155	1.32%
Georgia	698	6	-	703	0.15%
Guam	7	-	-	7	*
Hawaii	8,395	10	-	8,405	1.80%
Idaho	141	8	-	149	0.03%
Illinois	488	10	-	498	0.11%
Indiana	264	12	-	276	0.06%
Iowa	99	-	-	99	0.02%
Kansas	204	14	-	218	0.05%
Kentucky	109	7	-	116	0.02%
Louisiana	1,679	20	-	1,699	0.36%
Maine	454	1	-	455	0.10%
Maryland	10,568	5,284	-	15,852	3.39%
Massachusetts	4,893	93	-	4,986	1.07%
Michigan	2,229	24	-	2,252	0.48%
Minnesota	439	7	-	446	0.10%
Mississippi	1	s	-	1	*
Missouri	675	1	-	675	0.14%
Nebraska	58	17	-	75	0.02%

Table 11. (Continued)

State/Territory	Type			Total	Percent of U.S. Total
	Crystalline Silicon	Thin-Film	Concentrator		
Nevada	6,207	2,739	-	8,946	1.91%
New Hampshire	4,178	15	-	4,193	0.90%
New Jersey	35,916	5	-	35,921	7.69%
New Mexico	9,004	17	-	9,021	1.93%
New York	21,295	8,489	-	29,784	6.38%
North Carolina	1,099	12	-	1,112	0.24%
North Dakota	4	-	-	4	*
Ohio	3,738	354	-	4,092	0.88%
Oklahoma	240	21	-	261	0.06%
Oregon	2,188	28	-	2,216	0.47%
Pennsylvania	13,791	18	-	13,809	2.96%
Puerto Rico	109	-	-	109	0.02%
Rhode Island	1,186	-	-	1,186	0.25%
South Carolina	280	-	-	280	0.06%
South Dakota	57	1	-	58	0.01%
Tennessee	1,092	8	-	1,100	0.24%
Texas	10,403	1,247	-	11,650	2.49%
Utah	484	7	-	491	0.11%
Vermont	3,038	7	-	3,045	0.65%
Virgin Islands of the U.S.	4	-	-	4	*
Virginia	1,280	5	-	1,285	0.28%
Washington	2,318	35	-	2,353	0.50%
West Virginia	44	2	-	46	0.01%
Wisconsin	994	4	-	998	0.21%
Wyoming	339	6	-	345	0.07%
Total	431,562	35,603	-	467,165	100.00%

* = Less than 0.01 percent.

s = Value is less than 0.5 of the table metric; value is included in any associated total.

- = No data reported.

Note: Totals may not equal sum of components due to independent rounding.

Source: U.S. Energy Information Administration, Form EIA-63B, "Annual Photovoltaic Cell/Module Shipments Report."

Table 12. U.S. photovoltaic module shipments to commercial sector by State/territory and type, 2010

peak kilowatts					
State/Territory	Type			Total	Percent of U.S. Total
	Crystalline Silicon	Thin-Film	Concentrator		
Alabama	85	-	-	85	0.01%
Alaska	29	-	-	29	*
American Samoa	677	-	-	677	0.07%
Arizona	54,188	3,441	-	57,629	6.35%
Arkansas	-	475	-	475	0.05%
California	346,912	12,797	2,188	361,897	39.85%
Colorado	29,639	462	4	30,104	3.31%
Connecticut	5,696	1,145	-	6,841	0.75%
Delaware	4,251	1,140	-	5,391	0.59%
District of Columbia	140	-	-	140	0.02%
Florida	15,824	1,832	-	17,655	1.94%
Georgia	4,413	333	-	4,746	0.52%
Hawaii	8,707	697	-	9,404	1.04%
Idaho	109	27	-	136	0.01%
Illinois	2,941	4	-	2,945	0.32%
Indiana	2,092	1,166	-	3,258	0.36%
Iowa	10	4	-	14	*
Kansas	292	48	-	340	0.04%
Kentucky	72	134	-	206	0.02%
Louisiana	603	11	-	614	0.07%
Maine	160	3	-	163	0.02%
Maryland	15,689	166	-	15,856	1.75%
Massachusetts	23,952	208	-	24,160	2.66%
Michigan	4,143	995	-	5,137	0.57%
Minnesota	1,374	42	-	1,416	0.16%
Missouri	208	85	-	294	0.03%
Nebraska	33	24	-	57	0.01%
Nevada	10,262	796	-	11,059	1.22%
New Jersey	188,958	14,470	-	203,428	22.40%
New Mexico	10,520	62	-	10,582	1.17%
New York	24,029	831	-	24,860	2.74%
North Carolina	8,683	10	-	8,693	0.96%
North Dakota	21	-	-	21	*
Ohio	2,337	7,518	-	9,855	1.09%
Oklahoma	33	39	-	72	0.01%
Oregon	5,287	64	-	5,351	0.59%
Pennsylvania	30,845	4,512	-	35,357	3.89%
Puerto Rico	4,441	22	-	4,463	0.49%
Rhode Island	761	1	-	762	0.08%

Table 12. (Continued)

New Hampshire	272	26	-	298	0.03%
South Carolina	171	1	-	172	0.02%
South Dakota	20	-	-	20	*
Tennessee	4,280	44	-	4,324	0.48%
Texas	24,085	1,023	-	25,109	2.76%
Utah	820	3	-	823	0.09%
Vermont	7,350	1	-	7,351	0.81%
Virgin Islands of the U.S.	79	-	-	79	0.01%
Virginia	1,151	72	-	1,223	0.13%
Washington	1,292	56	-	1,348	0.15%
West Virginia	352	-	-	352	0.04%
Wisconsin	1,151	1,718	-	2,869	0.32%
Wyoming	85	-	-	85	0.01%
Total	849,526	56,506	2,192	908,224	100.00%

* = Less than 0.01 percent.

- = No data reported.

Note: Totals may not equal sum of components due to independent rounding.

Source: U.S. Energy Information Administration, Form EIA-63B, "Annual Photovoltaic Cell/Module Shipments Report."

Table 13. U.S. photovoltaic module shipments to industrial sector by State/territory and type, 2010

peak kilowatts					
State/Territory	**Type**			**Total**	**Percent of U.S.Total**
	Crystalline Silicon	**Thin-Film**	**Concentrator**		
Arizona	2,348	-	-	2,348	4.87%
California	30,080	-	924	31,004	64.31%
Colorado	2,713	-	-	2,713	5.63%
Delaware	200	-	-	200	0.41%
Florida	113	-	-	113	0.23%
Georgia	239	-	-	239	0.50%
Illinois	156	-	-	156	0.32%
Kentucky	501	-	-	501	1.04%
Louisiana	227	-	-	227	0.47%
Maryland	566	-	-	566	1.17%
Massachusetts	51	-	-	51	0.11%
Missouri	87	-	-	87	0.18%
Montana	98	-	-	98	0.20%
Nebraska	80	-	-	80	0.17%
New Jersey	4,538	-	-	4,538	9.41%
New Mexico	234	-	-	234	0.49%

State/Territory	Type			Total	Percent of U.S.Total
	Crystalline Silicon	Thin-Film	Concentrator		
New York	793	-	-	793	1.64%
North Dakota	90	-	-	90	0.19%
Oklahoma	406	-	-	406	0.84%
South Dakota	25	-	-	25	0.05%
Texas	1,795	66	77	1,938	4.02%
Utah	201	-	-	201	0.42%
Virginia	110	-	-	110	0.23%
Washington	997	-	-	997	2.07%
West Virginia	30	-	-	30	0.06%
Wyoming	464	-	-	464	0.96%
Total	47,141	66	1,001	48,208	100.00%

- = No data reported.

Note: Totals may not equal sum of components due to independent rounding.

Source: U.S. Energy Information Administration, Form EIA-63B, "Annual Photovoltaic Cell/Module Shipments Report."

Table 14. U.S. photovoltaic module shipments to electric power sector by State/territory and type, 2010

peak kilowatts					
State/Territory	Type			Total	Percent of U.S. Total
	Crystalline Silicon	Thin-Film	Concentrator		
Arizona	5,990	4	2,312	8,306	3.40%
California	12,427	26,519	-	38,946	15.96%
Colorado	20,059	-	1,001	21,060	8.63%
Florida	3,193	10,080	-	13,273	5.44%
Georgia	-	9	-	9	0.00%
Massachusetts	3,800	-	-	3,800	1.56%
Nevada	-	59,627	-	59,627	24.44%
New Jersey	2,134	500	-	2,634	1.08%
New Mexico	2,060	37,711	1,000	40,771	16.71%
North Carolina	19,389	-	-	19,389	7.95%
Ohio	-	9,154	-	9,154	3.75%
Pennsylvania	-	3,038	-	3,038	1.25%
Puerto Rico	4,000	-	-	4,000	1.64%
Texas	3,593	16,347	-	19,940	8.17%
Total	76,645	162,989	4,313	243,947	100.00%

- = No data reported.

Note: Totals may not equal sum of components due to independent rounding.

Source: U.S. Energy Information Administration, Form EIA-63B, "Annual Photovoltaic Cell/Module Shipments Report."

Table 15. U.S. photovoltaic module shipments to grid-connected centralized PV system by State/territory and type, 2010

peak kilowatts					
State/ Territory	Type			Total	Percent of U.S.Total
	Crystalline Silicon	Thin-Film	Concentrator		
Arizona	5,990	4	2,312	8,306	3.40%
California	12,427	26,519	-	38,946	15.96%
Colorado	20,059	-	1,001	21,060	8.63%
Florida	3,193	10,080	-	13,273	5.44%
Georgia	-	9	-	9	0.00%
Massachusetts	3,800	-	-	3,800	1.56%
Nevada	-	59,627	-	59,627	24.44%
New Jersey	2,134	500	-	2,634	1.08%
New Mexico	2,060	37,711	1,000	40,771	16.71%
North Carolina	19,389	-	-	19,389	7.95%
Ohio	-	9,154	-	9,154	3.75%
Pennsylvania	-	3,038	-	3,038	1.25%
Puerto Rico	4,000	-	-	4,000	1.64%
Texas	3,593	16,347	-	19,940	8.17%
Total	76,645	162,989	4,313	243,947	100.00%

- = No data reported.

Note: Totals may not equal sum of components due to independent rounding.

Source: U.S. Energy Information Administration, Form EIA-63B, "Annual Photovoltaic Cell/Module Shipments Report."

Table 16. U.S. photovoltaic module shipments to grid-connected distributed PV system by State/territory and type, 2010

peak kilowatts					
State/Territory	Type			Total	Percent of U.S. Total
	Crystalline Silicon	Thin-Film	Concentrator		
Alabama	242	-	-	242	0.02%
Alaska	43	-	-	43	*
American Samoa	677	-	-	677	0.05%
Arizona	99,666	3,470	-	103,135	7.32%
Arkansas	104	475	-	579	0.04%
California	583,068	29,212	3,112	615,393	43.69%
Colorado	46,026	665	4	46,695	3.32%
Connecticut	8,197	1,138	-	9,335	0.66%

State/Territory	Type			Total	Percent of U.S. Total
	Crystalline Silicon	Thin-Film	Concentrator		
Delaware	8,819	1,140	-	9,959	0.71%
District of Columbia	387	-	-	387	0.03%
Florida	21,928	1,639	-	23,567	1.67%
Georgia	5,107	315	-	5,422	0.38%
Guam	7	-	-	7	*
Hawaii	16,901	707	-	17,608	1.25%
Idaho	250	35	-	284	0.02%
Illinois	3,484	10	-	3,494	0.25%
Indiana	2,254	1,162	-	3,416	0.24%
Iowa	100	-	-	100	0.01%
Kansas	496	56	-	552	0.04%
Kentucky	618	141	-	759	0.05%
Louisiana	2,281	20	-	2,301	0.16%
Maine	510	1	-	511	0.04%
Maryland	26,117	5,421	-	31,537	2.24%
Massachusetts	28,894	294	-	29,189	2.07%
Michigan	6,370	1,006	-	7,377	0.52%
Minnesota	1,723	34	-	1,757	0.12%
Missouri	883	86	-	969	0.07%
Montana	358	10	-	368	0.03%
Nebraska	90	38	-	129	0.01%
Nevada	15,010	3,535	-	18,545	1.32%
New Hampshire	4,372	15	-	4,387	0.31%
New Jersey	229,374	14,435	-	243,810	17.31%
New Mexico	19,374	47	-	19,420	1.38%
New York	45,509	9,175	-	54,684	3.88%
North Carolina	9,781	19	-	9,800	0.70%
North Dakota	25	-	-	25	*
Ohio	6,074	7,851	-	13,925	0.99%
Oklahoma	273	43	-	316	0.02%
Oregon	7,471	92	-	7,563	0.54%
Pennsylvania	44,618	4,530	-	49,148	3.49%
Puerto Rico	4,550	22	-	4,572	0.32%
Rhode Island	1,947	-	-	1,947	0.14%
South Carolina	395	-	-	395	0.03%
South Dakota	77	1	-	78	0.01%
Tennessee	5,240	52	-	5,292	0.38%
Texas	34,004	2,176	77	36,257	2.57%

Table 16. (Continued)

State/Territory	Type			Total	Percent of U.S. Total
	Crystalline Silicon	Thin-Film	Concentrator		
Utah	1,292	7	-	1,299	0.09%
Vermont	10,388	7	-	10,395	0.74%
Virgin Islands of the U.S.	82	-	-	82	0.01%
Virginia	2,540	5	-	2,545	0.18%
Washington	3,588	87	-	3,675	0.26%
West Virginia	395	-	-	395	0.03%
Wisconsin	2,018	1,675	-	3,693	0.26%
Wyoming	421	-	-	421	0.03%
Total	1,314,422	90,848	3,193	1,408,462	100.00%

* = Less than 0.01 percent.

- = No data reported.

Note: Totals may not equal sum of components due to independent rounding.

Source: U.S. Energy Information Administration, Form EIA-63B, "Annual Photovoltaic Cell/Module Shipments Report."

Table 17. U.S. photovoltaic module shipments to off-grid Domestic PV system by State/territory and type, 2010

peak kilowatts					
State/Territory	Type			Total	Percent of U.S. Total
	Crystalline Silicon	Thin-Film	Concentrator		
Alabama	s	s	-	s	*
Alaska	55	-	-	55	0.51%
Arizona	1,055	-	-	1,055	9.64%
Arkansas	s	s	-	1	0.01%
California	2,032	200	-	2,232	20.40%
Colorado	230	-	-	230	2.10%
Connecticut	8	-	-	8	0.07%
Delaware	4	1	-	5	0.05%
District of Columbia	3	-	-	3	0.03%
Florida	96	-	-	96	0.88%
Georgia	242	-	-	242	2.21%
Hawaii	201	-	-	201	1.84%
Illinois	101	-	-	101	0.92%
Indiana	102	12	-	114	1.04%

State/Territory	Type			Total	Percent of U.S. Total
	Crystalline Silicon	Thin-Film	Concentrator		
Iowa	10	-	-	10	0.09%
Kentucky	64	-	-	64	0.59%
Louisiana	228	-	-	228	2.08%
Maine	104	-	-	104	0.95%
Maryland	573	-	-	573	5.24%
Massachusetts	1	-	-	1	0.01%
Michigan	1	-	-	1	0.01%
Mississippi	1	s	-	1	0.01%
Missouri	87	-	-	87	0.79%
Montana	119	-	-	119	1.09%
Nebraska	80	-	-	80	0.74%
Nevada	11	-	-	11	0.10%
New Hampshire	78	-	-	78	0.72%
New Jersey	38	-	-	38	0.34%
New Mexico	385	-	-	385	3.51%
New York	609	-	-	609	5.57%
North Carolina	2	-	-	2	0.02%
North Dakota	90	-	-	90	0.82%
Ohio	1	-	-	1	0.01%
Oklahoma	406	-	-	406	3.71%
Oregon	4	-	-	4	0.03%
Pennsylvania	18	-	-	18	0.17%
South Carolina	56	-	-	56	0.51%
South Dakota	25	-	-	25	0.23%
Tennessee	132	-	-	132	1.21%
Texas	1,612	-	-	1,612	14.73%
Utah	213	-	-	213	1.95%
Vermont	s	-	-	s	*
Virgin Islands of the U.S.	1	-	-	1	0.01%
Virginia	1	-	-	1	*
Washington	1,016	-	-	1,016	9.29%
West Virginia	31	2	-	33	0.30%
Wisconsin	127	-	-	127	1.16%
Wyoming	467	6	-	472	4.32%
Total	10,719	221	-	10,941	100.00%

* = Less than 0.01 percent.

s = Value is less than 0.5 of the table metric; value is included in any associated total.

- = No data reported.

Note: Totals may not equal sum of components due to independent rounding.

Source: U.S. Energy Information Administration, Form EIA-63B, "Annual Photovoltaic Cell/Module Shipments Report."

Table 18. U.S. photovoltaic module shipments to off-grid non-domestic PV system by State/territory and type, 2010

peak kilowatts					
State/Territory	Type			Total	Percent of U.S. Total
	Crystalline Silicon	Thin-Film	Concentrator		
Arizona	145	-	-	145	3.46%
California	502	111	-	613	14.62%
Colorado	s	29	-	29	0.70%
Connecticut	-	12	-	12	0.29%
Florida	-	260	-	260	6.20%
Georgia	-	24	-	24	0.57%
Illinois	-	4	-	4	0.10%
Indiana	-	4	-	4	0.10%
Iowa	-	4	-	4	0.10%
Kansas	-	6	-	6	0.14%
Louisiana	-	11	-	11	0.26%
Maine	-	3	-	3	0.07%
Maryland	133	30	-	163	3.89%
Massachusetts	-	7	-	7	0.17%
Michigan	-	12	-	12	0.29%
Minnesota	90	15	-	105	2.50%
Montana	98	-	-	98	2.33%
Nebraska	-	3	-	3	0.07%
Nevada	1,449	-	-	1,449	34.56%
New Hampshire	-	26	-	26	0.62%
New Jersey	-	39	-	39	0.93%
New Mexico	-	32	-	32	0.76%
New York	-	145	-	145	3.45%
North Carolina	-	3	-	3	0.07%
Ohio	-	21	-	21	0.49%
Oklahoma	-	17	-	17	0.41%
Rhode Island	-	1	-	1	0.02%
South Carolina	-	1	-	1	0.02%
Texas	667	160	-	827	19.73%
Utah	-	3	-	3	0.07%
Vermont	-	1	-	1	0.02%
Virginia	-	72	-	72	1.72%

State/Territory	Type			Total	Percent of U.S. Total
	Crystalline Silicon	Thin-Film	Concentrator		
Washington	3	3	-	6	0.14%
Wisconsin	-	47	-	47	1.13%
Total	3,088	1,106	-	4,193	100.00%

s = Value is less than 0.5 of the table metric; value is included in any associated total.

- = No data reported.

Note: Totals may not equal sum of components due to independent rounding.

Source: U.S. Energy Information Administration, Form EIA-63B, "Annual Photovoltaic Cell/Module Shipments Report."

End Notes

[1] DC stands for direct current, the type of power output by photovoltaic cells and modules. All capacity values in this report are in DC terms.

[2] Full-time equivalent is a unit measure of the number of employees on full-time schedules plus the number of employees on part-time schedules converted to a full-time basis.

INDEX

A

B

C

H

I

J

K

L

M

N

O

P

Q

R

S

T

U

V

W